M000266988

GIS for SCIENCE

APPLYING MAPPING AND SPATIAL ANALYTICS

DAWN J. WRIGHT AND CHRISTIAN HARDER, EDITORS

Esri Press | Redlands, California

Esri Press, 380 New York Street, Redlands, California 92373-8100
Copyright © 2019 Esri
All rights reserved

Printed in the United States of America
23 22 21 20 19   1 2 3 4 5 6 7 8 9 10

Christian Harder and Dawn J. Wright, eds.; *GIS for Science: Applying Mapping and Spatial Analytics*; DOI: https://doi.org/10.17128/9781589485303

Library of Congress Control Number: 2019936340

The information contained in this document is the exclusive property of Esri unless otherwise noted. This work is protected under United States copyright law and the copyright laws of the given countries of origin and applicable international laws, treaties, and/or conventions. No part of this work may be reproduced or transmitted in any form or by any means, electronic or mechanical, including photocopying or recording, or by any information storage or retrieval system, except as expressly permitted in writing by Esri. All requests should be sent to Attention: Contracts and Legal Services Manager, Esri, 380 New York Street, Redlands, California 92373-8100, USA.

The information contained in this document is subject to change without notice.

US Government Restricted/Limited Rights: Any software, documentation, and/or data delivered hereunder is subject to the terms of the License Agreement. The commercial license rights in the License Agreement strictly govern Licensee's use, reproduction, or disclosure of the software, data, and documentation. In no event shall the US Government acquire greater than RESTRICTED/LIMITED RIGHTS. At a minimum, use, duplication, or disclosure by the US Government is subject to restrictions as set forth in FAR §52.227-14 Alternates I, II, and III (DEC 2007); FAR §52.227-19(b) (DEC 2007) and/or FAR §12.211/12.212 (Commercial Technical Data/Computer Software); and DFARS §252.227-7015 (DEC 2011) (Technical Data – Commercial Items) and/or DFARS §227.7202 (Commercial Computer Software and Commercial Computer Software Documentation), as applicable. Contractor/Manufacturer is Esri, 380 New York Street, Redlands, CA 92373-8100, USA.

@esri.com, 3D Analyst, ACORN, Address Coder, ADF, AML, ArcAtlas, ArcCAD, ArcCatalog, ArcCOGO, ArcData, ArcDoc, ArcEdit, ArcEditor, ArcEurope, ArcExplorer, ArcExpress, ArcGIS, arcgis.com, ArcGlobe, ArcGrid, ArcIMS, ARC/INFO, ArcInfo, ArcInfo Librarian, ArcLessons, ArcLocation, ArcLogistics, ArcMap, ArcNetwork, ArcNews, ArcObjects, ArcOpen, ArcPad, ArcPlot, ArcPress, ArcPy, ArcReader, ArcScan, ArcScene, ArcSchool, ArcScripts, ArcSDE, ArcSdl, ArcSketch, ArcStorm, ArcSurvey, ArcTIN, ArcToolbox, ArcTools, ArcUSA, ArcUser, ArcView, ArcVoyager, ArcWatch, ArcWeb, ArcWorld, ArcXML, Atlas GIS, AtlasWare, Avenue, BAO, Business Analyst, Business Analyst Online, BusinessMAP, CityEngine, CommunityInfo, Database Integrator, DBI Kit, EDN, Esri, esri.com, Esri—Team GIS, Esri—The GIS Company, Esri—The GIS People, Esri—The GIS Software Leader, FormEdit, GeoCollector, Geographic Design System, Geography Matters, Geography Network, geographynetwork.com, Geoloqi, Geotrigger, GIS by Esri, gis.com, GISData Server, GIS Day, gisday.com, GIS for Everyone, JTX, MapIt, Maplex, MapObjects, MapStudio, ModelBuilder, MOLE, MPS—Atlas, PLTS, Rent-a-Tech, SDE, SML, Sourcebook•America, SpatiaLABS, Spatial Database Engine, StreetMap, Tapestry, the ARC/INFO logo, the ArcGIS logo, the ArcPad logo, the Esri globe logo, the Esri Press logo, The Geographic Advantage, The Geographic Approach, the GIS Day logo, the MapIt logo, The World's Leading Desktop GIS, Water Writes, and Your Personal Geographic Information System are trademarks, service marks, or registered marks of Esri in the United States, the European Community, or certain other jurisdictions. CityEngine is a registered trademark of Procedural AG and is distributed under license by Esri. Other companies and products or services mentioned herein may be trademarks, service marks, or registered marks of their respective mark owners.

Ask for Esri Press titles at your local bookstore or order by calling 800-447-9778, or shop online at esri.com/esripress. Outside the United States, contact your local Esri distributor or shop online at eurospanbookstore.com/esri.

Esri Press titles are distributed to the trade by the following:

In North America:
Ingram Publisher Services
Toll-free telephone: 800-648-3104
Toll-free fax: 800-838-1149
E-mail: customerservice@ingrampublisherservices.com

In the United Kingdom, Europe, Middle East and Africa, Asia, and Australia:
Eurospan Group
3 Henrietta Street
London WC2E 8LU
United Kingdom
Telephone: 44(0) 1767 604972
Fax: 44(0) 1767 601640
E-mail: eurospan@turpin-distribution.com

All images courtesy of Esri except as noted. Developmental editing by Mark Henry; Technology Showcase editing by Keith Mann; Cover layout and book design by Steve Pablo; Cover concept by John Nelson.

The cover image shows a lidar-derived colored hillshade of glacial ice and snowfields on Mount Rainier in Washington State, United States, compared to the glacial extent of the same area from 1924. The lidar data is from a survey completed in 2008 by the National Park Service. Image by Daniel Coe, Washington Geological Survey. The lidar portion of the image was modified from Robinson, Joel E., Thomas W. Sisson, and Darin D. Swinney. 2010. Digital Topographic Map Showing the Extents of Glacial Ice and Perennial Snowfields at Mount Rainier, Washington, based on the LIDAR survey of September 2007 to October 2008: US Geological Survey Data Series 549. The 1924 extent of glacial ice was derived from: US Geological Survey, 1924, Mount Rainier Quadrangle, Washington: US Geological Survey, 1 sheet, map scale 1:125,000.

See this book come alive at
GISforScience.com

PRAISE FOR *GIS FOR SCIENCE*

"This book is beautiful as well as illuminating, and it dramatizes the ways in which the new science of geospatial information is enriching and empowering all other scientific disciplines."

—James Fallows, staff writer, *The Atlantic*; former chief speechwriter for President Jimmy Carter

"If you love maps like I do, you'll be drawn to this book. But you'll quickly discover so much more: the power of harnessing multiple perspectives and data types that infuse maps with even more meaning and catalyze new insights. A veritable treasure trove of ideas. "

—Jane Lubchenco, environmental scientist, marine ecologist, former Administrator of the National Oceanic and Atmospheric Administration (2009-2013); former U.S. Science Envoy for the Ocean (2014-2016); university distinguished professor

"GIS has become *the* foundational tool for all things environmental—from conservation to climate change to environmental justice. This astonishing book beautifully displays GIS in all its scientific, artistic, and creative splendor."

—Peter Kareiva, director, UCLA Institute of the Environment and Sustainability

"Dawn Wright and Christian Harder have given us a geoscience book for the twenty-first century! Cutting-edge research examples and gloriously illustrated state-of-the-art, GIS-enabled techniques come together to show us how to understand our planet in ways not possible even a few years ago."

—Margaret Leinen, Director of Scripps Institution of Oceanography and UC San Diego Vice Chancellor for Marine Sciences

"The only thing changing faster than Earth's environment and our species' imprint on it, for better and worse, is the information environment. In that noisy realm, trolls and other troublemakers get the headlines. But this essential and beautiful book illuminates how a host of innovators are gleaning meaning from data and helping shape a sustainable human journey."

—Andrew Revkin, strategic adviser, National Geographic Society, and coauthor of *Weather: An Illustrated History, from Cloud Atlases to Climate Change*

"The Science of Where® comes alive in **GIS for Science**. The book is, yes, informative, helping us understand how the world works, how it looks, and how we see it through images, maps, and more. Above all, it is dazzling, combining knowledge with a sense of wonder, bringing a desire to press for more discovery, and invoking a deep appreciation for why smart decisions spring from taking science to action."

—Lynn Scarlett, vice president, The Nature Conservancy; chair, Science Advisory Board, NOAA

"A textbook and a work of art."

—Len Kne, U-Spatial associate director,
University of Minnesota Research Computing, Office of the Vice President for Research

"Illustrating the power of geospatial analytics to address pressing challenges facing our planet, *GIS for Science* is a *tour de force*. The editors and contributors have produced a visual delight that will inspire and enlighten researchers, citizen scientists, and the public about the contribution of the geographic perspective to the scientific process."

—Sergio Rey, director, Center for Geospatial Sciences, UC Riverside

"The editors and authors of this unique graphical science book, published by Esri Press, show the criticality of asking the 'where' question when looking for answers to the 'why' question. It is visually stunning and will certainly lead to an expanded cohort of citizen scientists."

—Noel Cressie, distinguished professor, University of Wollongong, Australia

"Given the relevance of this geospatial perspective for all aspects of society, I hope this beautiful book will inspire a wide range of people to embrace The Science of Where®."

—Thomas Crowther, director, Global Forest Biodiversity Initiative, ETH-Zurich

"With vivid imagery, lucid writing, interactive learning, and compelling, relevant examples from Earth's past, present, and future, *GIS for Science* is a modern manual for understanding that integrative spatial analysis and visualization is the big data revolution most vital to the quality of all life on Earth."

—Healy Hamilton, chief scientist, NatureServe

"There is no better tool to understand our place in the world than GIS, and this book puts its power on beautiful display. It's a book for scientists and all of Earth's stewards."

—Jessica Hellmann, director, Institute on the Environment, University of Minnesota

"This beautifully illustrated and inspiring book brings home the power of today's technology with unique effectiveness, telling and illustrating stories from the earth sciences in novel and powerful ways. A must-have book for anyone concerned about the planet's future."

—Mike Goodchild, distinguished emeritus professor and research professor of geography, UC Santa Barbara

CONTENTS

Introduction

GIS for Science: A Framework and a Process—Jack Dangermond and Dawn J. Wright, Esri · viii
Introduction by the Editors—Dawn J. Wright and Christian Harder, Esri · ix
Reflections on a Blue Marble: An Astronaut's View—Kathryn Sullivan (ret.), NOAA · x

Part 1: How Earth Works · 2

Global Ecosystem Mapping—Roger Sayre, US Geological Survey · 4

Using advanced geospatial technology, a team of public- and private-sector scientists have created a high-resolution, standardized, and data-derived map of the world's ecosystems—a global dataset useful for studying the impacts of climate change, as well as the economic and noneconomic value these ecosystems provide.

What Lies Beneath—Daniel Coe, Washington Geological Survey · 22

For scientists studying landslides and other natural hazards in the geologically active state of Washington, lidar imagery has become an invaluable new data resource that enables one to literally see Earth's surface, even in places where trees and vegetation obscure the landscape.

The Anatomy of Supervolcanoes—Melanie Brandmeier, Esri Germany · 42

Working in the shadows of some of the most remote volcanic regions on the planet, geologists use geostatistical analysis to reveal the space-time patterns of volcanic super-eruption in the Central Andes of South America.

Predicting Global Seagrass Habitats—Orhun Aydin and Kevin A. Butler, Esri · 58

Using machine-learning techniques to study a mostly hidden but environmentally crucial marine resource, scientists are building geographically linked models that show where seagrasses are expected to flourish under differing ocean conditions.

Part 2: How Earth Looks · 70

Extreme Heat Events in a Changing Climate—Olga Wilhelmi and Jennifer Boehnert, NCAR · 72

Extreme heat is a major public health concern, and in response, scientists are using GIS to aid public officials in monitoring the frequency and intensity of forthcoming extreme heat events.

Finding a Way Home—Lauren Griffin and Este Geraghty, Esri · 84

This chapter presents a glimpse into the homelessness crisis taking place across America and describes how GIS can help cities, agencies, and spatial analysts understand, prevent, and manage this human dilemma.

Restoring Coastal Marine Habitats—Zach Ferdaña, Laura Flessner, Matt Silveira, and Morgan Chow, The Nature Conservancy; Tom Brouwer, FloodTags; and Omar Abou-Samra, American Red Cross · 104

Mapping the bond between people and nature, scientists are using geospatial technologies to build coastal resilience by addressing rising sea levels and other impacts of climate change.

Modeling Bird Responses to Climate Change—Molly Bennet, with Brooke Bateman, David Curson, Gary Langham, Curtis Smalling, Lotem Taylor, Chad Wilsey, and Joanna Wu, National Audubon Society · 118

Using geospatial analysis and mapping tools, a century-old conservation group is targeting which habitats will be most critical for birds in a warmer world, telling stories with maps to show bird lovers just what is at stake and how they can help protect the places that birds and people need to thrive.

Part 3: How We Look at Earth 140

Mapping Ancient Landscapes—Jason Ur and Jeffrey Blossom, Harvard University 142
Racing against the clock as development encroaches on important Kurdish heritage sites, a team of landscape archaeologists deploys drones and comparative image analysis to capture previously undetected ancient settlements.

Identifying the Natural Efficient Frontier—Jeff Allenby, Chesapeake Conservancy; and Lucas Joppa and Nebojsa Jojic, Microsoft Research 166
To improve conservation efforts across the entire US, scientists are leveraging artificial intelligence and satellite imagery within GIS across large landscapes to find the very best places for restoration.

Part 4: Training Future Generations of Scientists 180

A Glacier in Retreat—Jacki Klancher, Todd Guenther, and Darran Wells, Central Wyoming College 182
Wyoming is the third-most glaciated state in the United States after Alaska and Washington. The quest to measure the extent of ice retreat and predict the implications of losing the state's 80-plus glaciers has led a multidisciplinary research team to the Dinwoody Glacier at the base of Gannett Peak—Wyoming's tallest mountain.

Panamapping: GIS for Conservation Science—Dan Klooster, David Smith, Nathan Strout, University of Redlands; Experience Mamoní; and Fundación Geoversity 200
Geographic information system (GIS) technology supports conservation goals in Panama by revealing how physical features of the landscape interact with current and historical human uses of the land, allowing conservation managers to visualize and communicate processes of forest change, locate critical areas, and plan conservation activities.

Part 5: Technology Showcase 214

Emergence of the Geospatial Cloud 216

Equal Earth Projection 218

Science of the Hex 221

Modeling the Footprint of Human Settlement 222

Modeling Green Infrastructure 224

Jupyter™ Notebook Analysis 226

3D Empirical Bayesian Kriging 228

National Water Model 230

A High-Resolution Martian Database 233

Sentinel-2 Imagery Viewer 234

The Power of Storytelling for Science 236

GIS FOR SCIENCE: A FRAMEWORK AND A PROCESS

by Jack Dangermond, *Founder and President, Esri*
and Dawn J. Wright, *Chief Scientist, Esri*

Science—that wonderful endeavor in which someone investigates a question or a problem using reliable, verifiable methods and then broadly shares the result, has always been about increasing our understanding of the world. In the beginning, we applied geographic information systems (GIS) to science—to biology, ecology, economics, or any of the other social sciences. It wasn't until around 1993, when Professor Michael Goodchild coined the term *GIScience,* that the world began to realize that GIS is a science in its own right. Today, we call this The Science of Where®. GIS incorporates sciences such as geology, data science, computer science, statistics, humanities, medicine, decision-support science, and much more. It integrates all these disciplines into a kind of metascience, providing a framework for applying science to almost everything, merging the rigor of the scientific method with the technologies of GIS. The study of where things happen, it turns out, has great relevance.

So why is this work all so important right now? We live in a world that faces more and more challenges. We see, we hear, and we read daily about such issues as growing population (some would say overpopulation), climate change, loss of nature, loss of biodiversity, social conflicts, urbanization, natural disasters, pollution, and political polarization. We also confront the realities of food, water, and energy shortages, and general overconsumption of resources. These concerns are not trivial for the individuals and organizations working in these fields. We must do everything we can to better understand these crucial issues and form better collaborations to address the challenges.

Our world at the same time is undergoing a massive digital transformation. Science always has been about increasing our understanding of the world. But it is also about using that understanding to enable innovation and transformation. It is about what we can measure, how we analyze things, what predictions we make, how we plan, how we design, how we evaluate, and ultimately, how we weave it all together in a kind of fabric across the planet.

What GIS provides is a language to help us understand and manage inside, between, and among organizations, to positively affect the future of the planet. It is also a framework in which we can compile and organize maps, data, and applications. We can visualize and analyze the relationships and patterns among our datasets, perform predictive analytics, design and plan with the data, and ultimately transform our thinking into action to create a more sustainable future. This technology also delivers a new way to empower people to easily use spatial information. As Richard Saul Wurman has said, "Understanding precedes action." Esri is driven by the idea that GIS as a technology is the best way to address the challenges of today and the future.

> Science itself is driven by the organic human instinct to dream, to discover, to understand, to create.

This book is full of examples that show how GIS advances rigorous scientific research. It shows how many science-based organizations use ArcGIS as a comprehensive geospatial platform to support spatial analysis and visualization, open data distribution, and communicate. In some cases, we use this research to preserve and restore iconic pieces of nature—revered and sacred places worthy of being set aside for future generations. These places belong to nature, and they also belong to science.

As scientists, the discipline of the scientific process is the central organizing principle of our work. But science itself is also driven by the organic human instinct to dream, to discover, to understand, to create. The Science of Where is a concept that brings these impulses together as we seek to transform the world through maps and analytics, connecting everyone, everywhere, every day through science. At Esri, we can't wait to see what you and your colleagues will achieve with geospatial technology.

INTRODUCTION BY THE EDITORS

This book is about science and the scientists who use GIS technology in their work. This contributed volume is for professional scientists, the swelling ranks of citizen scientists, and anyone interested in science and geography. Our world, now two decades into the twenty-first century, seems to be entering a crucial time in history in which humanity still can create a sustainable future and a livable environment for all life on the planet. But if we look critically at the facts, no informed observer can refute the reality that the current downward trajectory does not bode well.

Our first objective in assembling this volume was to select relevant and interesting stories about the state of the planet in 2019. We looked for a cross section of sciences and scientists studying a wide range of problems.

GIS has found its way into virtually all the sciences, but the reader will notice that earth and atmospheric sciences are especially well represented. Web GIS patterns and a simultaneous explosion of earth-observation sensors fuel this growth. Between all the satellites, aircraft, drones, and myriad ground-based and tracking sensors, the science community is now awash in data. Well-integrated GIS solutions integrate all this big data into a common operating platform—a digital, high-resolution, multiscale, multispectral model of our world.

Despite all these advances, science is under attack on many fronts. From fake news to political pressure, science is too often being used as a political tool at a time when level-headed, objective scientific thinking is required. We are convinced that GIS offers a unique platform for scientists to elevate their work above the fray. We invite you to read these stories in any order; the common thread is that all this work happens at the intersection of GIS and science. As you read through these stories, you'll see that GIS is a cross-cutting, enabling technology, whose use is limited only by our imaginations.

In some cases, like the fascinating work of the US Geological Survey in developing global ecosystem characterizations of the land and ocean, GIS and spatial analysis are at the core of the science. These innovations in science could only happen in the context of an advanced GIS. In other cases, like the story of glaciologists using ground-penetrating radar to measure ice loss in the high-country glaciers of Wyoming, GIS embeds itself in the science but is still mission-critical in terms of expedition planning, backcountry navigation, and analysis. GIS also serves as a vital storytelling platform that brings critical details of important research to stakeholders in the local community.

How the book and website work together

It's impossible to describe the full breadth and scope of what GIS means for science and scientists without showing digital examples. So we have created a companion and complement to this book online. You can access it here:

GISforScience.com

This unique website, comprising collections of ArcGIS® StoryMaps™ stories, apps, and digital maps, brings the real-world examples to life and demonstrates the storytelling power of the ArcGIS® platform. The website also includes links to learning pathways from the Learn ArcGIS site (Learn.ArcGIS.com) and blogs related to the practical use of ArcGIS in each of the case studies.

REFLECTIONS ON A BLUE MARBLE

AN ASTRONAUT'S VIEW, BY KATHRYN SULLIVAN

From the observation windows of the space shuttle *Atlantis*, you see 16 sunrises and sunsets every day. The visual experience can be confusing. But when the shuttle crosses over the terminator into the dark side, for a couple of minutes you are high enough up that you can look down to the dark earth and start to see lights appear. There, people had just watched the sunset. Maybe they were having dinner or coming in from their outdoor activities. I could imagine someone on the ground looking up, some little kid, saying, "Look! There goes a satellite!" And those people were peering right at us. It is one of those amazing, perspective-shifting experiences that can remind us how singular, how unitary, how "one" this place is that we live upon.

This experience is new in our lifetime. We are still the first generation of human beings to have the ability to see and measure the planet instantaneously. To see all at once and to understand the planet, to put it on a map, to look at, study, and appreciate the impact of these various aspects of Earth is unprecedented. The "Blue Marble." The vegetation data. Biosphere on land and in the ocean. Soil moisture. Winds. Ocean-surface temperature. The immensity boggles the mind. To see simultaneous snapshots of these conditions across the entire planet is a miracle in itself, but when you capture and analyze that torrent of data in a GIS, you suddenly get deep and actionable meaning. This data is inherently geospatial, which is why my former agency, the National Oceanic and Atmospheric Administration (NOAA), and so many other scientific organizations around the planet rely on GIS software tools to get work done.

Nowadays, this capability to know where and what and when things occur on our planet gives us a remarkable predictive capability. This information gives us foresight to minimize damage, protect physical property, and, most importantly, save lives. And it's not just violent weather, volcanic phenomena, and other fast-onset events that affect what we do and how we live on this planet. It is increasingly clear that our species also must deal with the slow, creeping, direct challenges of planetary change.

This scale of situational awareness about our planet did not exist in the history of humankind before the Space Age. We are just now beginning to fully live with and learn how to use this kind of omniscience. We are just learning how to implement our knowledge into the way we live on this planet. Maps, particularly online storytelling maps, create access for nonscientists to cognitively and emotionally and intellectually grasp the message. All this data collection and analysis that we can now do so easily is nearly meaningless if we can't tell interesting and compelling stories about our scientific work.

This nighttime image shows US city lights in at least half a dozen southern states from some 225 miles above Earth. Visible are lights from areas in the Gulf Coast states of Texas, Louisiana, Mississippi, and Alabama.

The power of a map is that it can put time and place and phenomena together and send that information to our brains through the most potent input sensor human beings have—our eyes. This accelerator is remarkable for the comprehension, engagement, and use of the data that tells us where things are happening on our planet. How is our planet changing through time and space? And how are those changes intersecting my life, my business, my community, and my country? Because when you get out of the nice, cool model geometries and return from the ethereal inspirational perspective of an astronaut, you find that all of us living on this dynamic planet have had an undeniable impact. And now we can readily see the downside in terms of resource depletion, loss of open space, pollution, and so on.

This new perspective afforded by the Space Age, coupled with the spatial and temporal ability that mapmaking gives us, is the one powerful tool we have that can bring these longer-time changes into focus. Today, we have an opportunity to use this technology to inform our decisions—to help us live more wisely and well on this planet.

Dr. Kathryn Sullivan flew on three space shuttle missions as a geophysical scientist and was administrator of NOAA from 2012 to 2016. (NASA)

Photo by Reid Wiseman/NASA

You can see it just below Saturn's rings—Carl Sagan's pale, blue dot. For his land-mark 1970s PBS special *Cosmos*, Sagan and the creative people at PBS put together wonderful animations where you watch all the other planets go by as an imaginary spacecraft zooms away from Earth and into the universe to look from afar at this blue dot that we call home. Carl was a scientist and a poet and an amazing person. He would be awed but not surprised to learn that we have these views not just as animations but for real now. This image is from the NASA *Cassini* spacecraft. At the time of Carl's television show, we human beings were already a space-traveling species. And we have kept pressing on, exploring anew, and stepping back from our planet, looking outward, trying to understand the whole, trying to understand all the various worlds.

Earth

This image from the Cassini spacecraft was taken just weeks before it plunged into Saturn in late 2017, more than 20 years after its launch. That image is our pale, blue planet seen through the rings of Saturn. You and everyone you know are in this picture. (Cassini Imaging Team)

GIS for SCIENCE

PART 1
HOW EARTH WORKS

This section includes examples of GIS helping scientists to gain better insight and understanding of Earth process and function in natural science fields such as geology, ecology, oceanography, climatology, cryospheric science, and conservation biology. By way of reliable, verifiable spatial analysis and visualization, GIS helps physical scientists answer a myriad of questions about spatial patterns in the natural environment (geosphere, biosphere, hydrosphere, atmosphere) and what process is responsible for those patterns. GIS is also a modern platform for the open sharing of data and for compelling science communication at a multiple of scales (e.g., individual researcher, lab workgroup, multi-department, multi-university, university-to-agency collaboration, and citizen engagement).

The Elwahs River's watershed in Washington State. Hundreds of thousands of salmon swam in the Elwha's pristine waters until the early twentieth century. Loggers harvesting the region's rich, old-growth forests eyed the Elwha—with its rushing waters and narrow canyon—as an ideal source of hydropower.

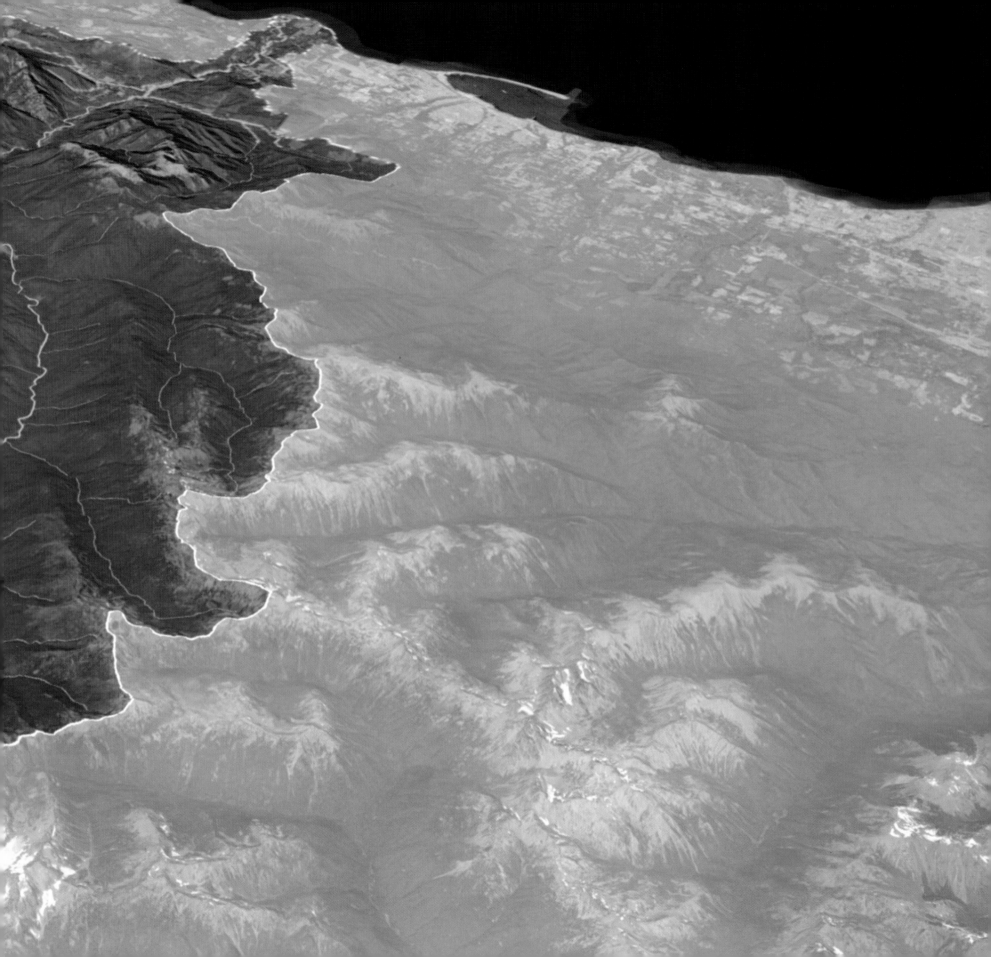

GLOBAL ECOSYSTEM MAPPING

Using advanced geospatial technology, a team of public- and private-sector scientists have created a high-resolution, standardized, and data-derived map of the world's ecosystems—a global dataset useful for studying the impacts of climate change, as well as the economic and noneconomic value these ecosystems provide.

By Roger Sayre, US Geological Survey

This map of the world synthesizes data on climate, landforms, geology, and vegetation to capture the total variety of ecological land units (ELUs) on the planet's land masses. There were 3,639 distinct land units identified.

INTRODUCTION

Anyone who is seriously interested in geography is probably familiar with an unassuming, spiral-bound, calendar-format book called *Geography for Life: National Geography Standards*,[1] now in its second edition. It contains the geography curriculum for students in the fourth, eighth, and twelfth grades and is a valuable read for anyone, educators and noneducators alike, who wants to know what the geographically informed person needs to know. The book opens with an emphasis on two fundamental geographic perspectives that can and should always be taken into account when thinking about life and the earth: the spatial perspective and the ecological perspective.

It is interesting and refreshing to finally see these two geographic perspectives as primary. To have those perspectives spelled out so clearly in the *National Geography Standards* is powerful and affirmational to the growing ranks of spatial ecologists trying to apply "The Science of Where" to "The Ecology of Earth." In that grand and evolving synthesis, there are many important questions to be asked, many thorny problems to be tackled, and fortunately, many opportunities to make a difference for the betterment of society and the planet. One of those opportunities is the conservation and sustainable development of our global ecosystems.

Ecosystems give us, as humans, goods (e.g., food, water, fiber, fuel, etc.) and services (e.g., water quality maintenance, flood control, carbon sequestration) that are critical for our survival. If ecosystems are allowed to persist on the planet, they will continue to provide those goods and services to future generations. This existential dependence of humans on natural ecosystems is recognized in the Sustainable Development Goals (SDGs)[2] that global leaders adopted in 2015 as part of the United Nations 2030 Agenda for Sustainable Development. Three of the 17 SDGs focus specifically on the protection of different kinds of terrestrial ecosystems (SDG 15), coastal and marine ecosystems (SDG 14), and freshwater ecosystems (SDG 6).

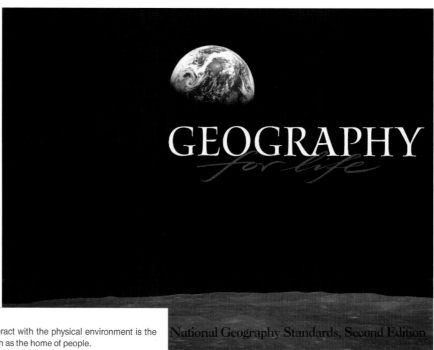

National Geography Standards, Second Edition

This well-organized set of curriculum standards is designed to stimulate the geographic imagination of students and is a practical resource for directing teachers how to present various issues in the field of geography. The vast majority of states have incorporated parts or all of the 18 geography standards into their state standards. Geography for Life has sold nearly 100,000 copies, and numerous textbooks have incorporated the standards into lesson plans and exercises.

Looking at the World in Multiple Ways: Geographic Perspectives

Where something occurs is the spatial perspective; how life forms interact with the physical environment is the ecological perspective. We need both perspectives to comprehend Earth as the home of people.

Perspectives, knowledge, and skills comprise the content of geography. In general, a perspective is a framework that can be used to interpret the meanings of experiences, events, places, persons, cultures, and physical environments. Having a perspective means looking at our world through a lens shaped by personal experience, selective information, and subjective evaluation. The perspectives and the questions to which they lead distinguish geography from other approaches, such as historic or economic. A perspective provides a frame of reference for asking and answering questions, identifying and solving problems, and evaluating the consequences of alternative actions.

It is essential to be aware that many different perspectives exist and that learning to understand the world from many points of view enhances our knowledge and skills. It is also essential to realize that our perspectives incorporate all life experiences and draw upon knowledge from many fields of inquiry. Therefore, people cannot be neatly boxed into specific categories based on their cultural experiences, ethnic backgrounds, age, gender, or any other life-status characteristic.

Acquiring, understanding, and using a wide variety of perspectives are essential to becoming a geographically informed person. Such a person knows that each individual has personal points of view based in unique life experiences; accepts the existence of diverse ways of looking at the world; understands how different perspectives develop; is aware that perspectives incorporate values, attitudes, and beliefs; considers a range of perspectives when analyzing, evaluating, and solving a problem; and understands that perspectives are subject to change.

Although the field of geography includes many different perspectives, geographers depend upon two perspectives in particular to frame their understanding of people and places in the world—the spatial perspective and the ecological perspective.

The Spatial Perspective

A historical perspective focuses on the temporal dimension of human experience (time and chronology), while geography is concerned with the spatial dimension of human experience (space and place). The space of Earth's surface is the fundamental characteristic underpinning geography. The essential issue of *whereness*—embodied in specific questions such as, "Where is it? Why is it there?"—helps humans contemplate the context of spatial relationships in which the human story is played out. Understanding spatial patterns and processes is essential to appreciating how people live on Earth. People who approach knowing and doing with a habit of inquiring about whereness possess a spatial perspective.

The Ecological Perspective

Earth is composed of living and nonliving elements interacting in complex webs of ecological relationships that occur at multiple levels. Humans are part of the interactive and interdependent relationships in ecosystems and are one among many species that constitute the living part of Earth. Human actions modify physical environments and the viability of ecosystems at local to global scales. The survival of humans and other species requires a viable global ecosystem. Understanding Earth as a complex set of interactive living and nonliving elements is fundamental to knowing that human societies depend on diverse small and large ecosystems for food, water, and all other resources. People who regularly inquire about connections and relationships among life forms, ecosystems, and human societies possess an ecological perspective.

Understanding and using the spatial and ecological perspectives helps geographers understand how to interpret nature and societies on Earth. Viewed together, the geographic perspective overall encompasses an understanding of spatial patterns and processes on Earth and its web of living and nonliving elements interacting in complex webs of relationships within nature and between nature and societies.

A fully developed geographic perspective, therefore, involves an integration of both spatial and ecological points of view, as well as a consideration of other related perspectives that may be useful in understanding and interpreting the world.

THE NEED FOR GLOBAL ECOSYSTEMS MAPS

SUSTAINABLE DEVELOPMENT G⚙ALS

1 NO POVERTY	2 ZERO HUNGER	3 GOOD HEALTH AND WELL-BEING	4 QUALITY EDUCATION	5 GENDER EQUALITY	6 CLEAN WATER AND SANITATION
7 AFFORDABLE AND CLEAN ENERGY	8 DECENT WORK AND ECONOMIC GROWTH	9 INDUSTRY, INNOVATION AND INFRASTRUCTURE	10 REDUCED INEQUALITIES	11 SUSTAINABLE CITIES AND COMMUNITIES	12 RESPONSIBLE CONSUMPTION AND PRODUCTION
13 CLIMATE ACTION	14 LIFE BELOW WATER	15 LIFE ON LAND	16 PEACE, JUSTICE AND STRONG INSTITUTIONS	17 PARTNERSHIPS FOR THE GOALS	SUSTAINABLE DEVELOPMENT GOALS

The Sustainable Development Goals, otherwise known as the Global Goals, are a universal call to action to end poverty, protect the planet, and ensure that all people enjoy peace and prosperity. The crucial need for humans to improve the conservation and sustainable development of global ecosystems is recognized in several of the SDGs that countries have adopted through a United Nations resolution.

Freshwater: by 2020, protect and restore water-related ecosystems, including mountains, forests, wetlands, rivers, aquifers, and lakes.

Marine: by 2020, sustainably manage and protect marine and coastal ecosystems to avoid significant adverse impacts— including by strengthening their resilience—and take action for their restoration to achieve healthy and productive oceans. By 2020, conserve at least 10 percent of coastal and marine areas, consistent with national and international law, on the basis of the best available scientific information.

Terrestrial: by 2020, ensure the conservation, restoration, and sustainable use of terrestrial and inland freshwater ecosystems and their services, in particular, forests, wetlands, mountains, and drylands. By 2030, ensure the conservation of mountain ecosystems, including their biodiversity, in order to enhance their capacity to provide benefits that are essential for sustainable development.

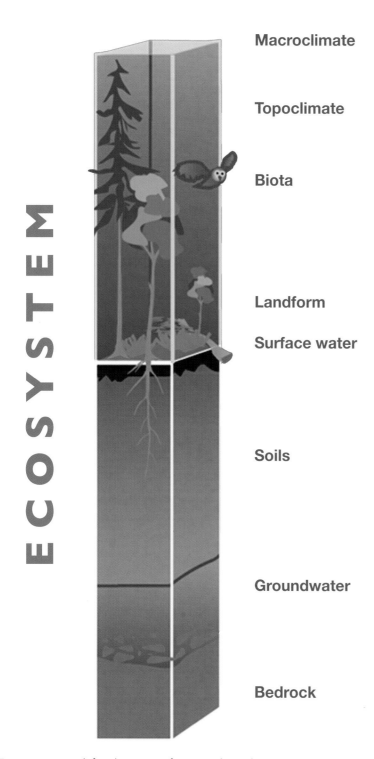

ECOSYSTEM

Macroclimate

Topoclimate

Biota

Landform

Surface water

Soils

Groundwater

Bedrock

Ecosystems are defined as areas of unique physical environment settings and biological assemblages. We can envision an ecosystem at any given point on Earth's surface as a vertical integration of its structural elements. So a terrestrial ecosystem is defined, for example, by its combination of climate regime, land-forms, organisms, substrate, and so on (the ecological perspective).[3]

It is clear by UN mandate that the need to protect ecosystems from threats is key to the sustainability of human society. It is not so clear, however, what ecosystems are, where they are located on the planet, what condition they are currently in, what may be threatening them, and how vulnerable they are to those threats. The consequences of failing to protect these ecosystems are also unclear. Understanding ecosystems is a challenge. Current and future generations of students in the United States will be increasingly knowledgeable about ecosystems, thanks in part to National Geography Standard No. 8 on "the characteristics and spatial distribution of ecosystems and biomes on Earth's surface."

However, ecosystems are fragile, and they are disappearing from the planet at unprecedented rates.[4] That fragility is one reason the Group on Earth Observations (GEO,[5] an intergovernmental consortium of about 200 nations and organizations) has commissioned the development of standardized, robust, and management-practical global ecosystem maps for terrestrial, freshwater, and marine environments. The US Geological Survey (USGS), with its institutional prioritization of the ecological perspective as a foundation of its overall science strategy,[6] was asked to lead that task and quickly turned to Esri for support with the spatial perspective.

TASK: Develop a standardized, robust, and practical global ecosystems classification and map for the planet's terrestrial, freshwater, and marine ecosystems, along with a web-enabled framework of data, tools, and work-flows that will be used to create and publish authoritative physiographic and ecological land classifications of the earth's surface at several scales.

—from GEOSS Task GI-14: Global Ecosystem Mapping

The USGS and Esri joined ecosystem experts from around the world to rapidly advance the science of ecosystems geography. In a successful and productive public, private, academic, and nongovernmental organization (NGO) partnership, the Global Ecosystem Mapping (GEM) team produced several first-of-their-kind, high-spatial-resolution, globally comprehensive ecosystem maps for terrestrial, freshwater, and marine domains. The team is approaching this task with the simple premise that Earth is indeed a multifaceted gem of sorts (in the shape of an oblate spheroid, of course) and that the facets of the gem represent ecosystems on the land and sea surfaces. Now known as "facet mapping," this innovative approach breaks the environment down into the smallest mappable terrestrial ecosystems, called *ecological facets*, or EFs.

MAPPING APPROACH

The analogy of ecosystems as gem facets breaks down when considering uniqueness. The facets of an actual gemstone need to be identical and geometrically perfect, and dissimilar facets may render the gem flawed. Ecosystem facets, on the other hand, are distinguished from one another by their differences in pattern. We distinguish ecosystems based on their unique physical environment settings and biological assemblages, and how they change over time. We envision an ecosystem at any given point in time as a vertical integration of its structural elements. So a terrestrial ecosystem is defined, for example, by its combination of climate regime, landforms, organisms, substrate, and so on (the ecological perspective).

As the structural elements of ecosystems change over space (the spatial perspective), so then does the ecosystem itself change over space. For anyone familiar with the basic workings of DNA, this idea is a little like taking a Mendelian genetic approach to ecosystems and thinking of ecosystem variation as an expression of a kind of "ecological DNA." Global ELUs were mapped as distinct combinations of climate, landform, lithology, and land cover. These four inputs were each mapped for the planet as individual layers, which were then stacked up in a GIS and spatially combined into one global data layer at a spatial resolution of 250 m.

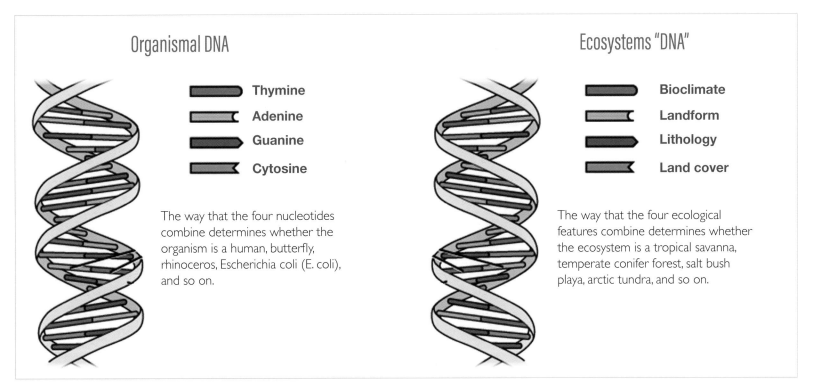

Organismal DNA

- Thymine
- Adenine
- Guanine
- Cytosine

The way that the four nucleotides combine determines whether the organism is a human, butterfly, rhinoceros, Escherichia coli (E. coli), and so on.

Ecosystems "DNA"

- Bioclimate
- Landform
- Lithology
- Land cover

The way that the four ecological features combine determines whether the ecosystem is a tropical savanna, temperate conifer forest, salt bush playa, arctic tundra, and so on.

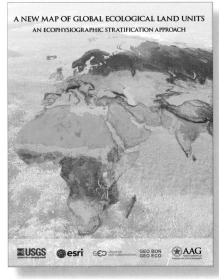

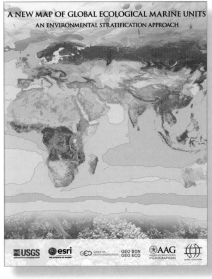

The GEM team has mapped ecological land units using the approach of identifying and mapping unique physical environments and their associated biological assemblages as distinct ecosystems. The published ELU work[7] and the later ecological marine unit (EMU) work[8] were scientifically rigorous, globally comprehensive, and high spatial resolution, producing state-of-the-art characterizations of global ecological environments.

ECOLOGICAL LAND UNITS

This analysis produced a staggering number (106,959) of ecological facets, each one a distinct combination of classes from each of the four input layers. To reduce this complexity, the 100,000+ facets were generalized into 3,639 ELUs. The map of global ELUs was presented using an "orange peel" (Goode's homolosine) map projection, and advanced cartographic techniques made the terrain visually palpable with colors that reflect wet/dry and hot/cold climates.

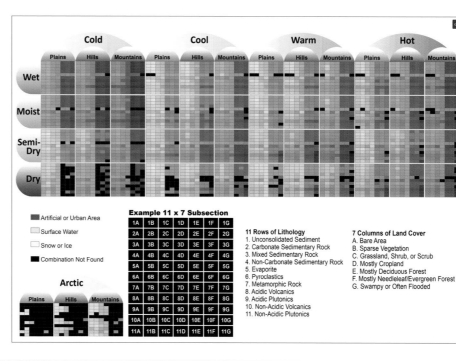

	Artificial or Urban Area
	Surface Water
	Snow or Ice
	Combination Not Found

Example 11 x 7 Subsection

1A	1B	1C	1D	1E	1F	1G
2A	2B	2C	2D	2E	2F	2G
3A	3B	3C	3D	3E	3F	3G
4A	4B	4C	4D	4E	4F	4G
5A	5B	5C	5D	5E	5F	5G
6A	6B	6C	6D	6E	6F	6G
7A	7B	7C	7D	7E	7F	7G
8A	8B	8C	8D	8E	8F	8G
9A	9B	9C	9D	9E	9F	9G
10A	10B	10C	10D	10E	10F	10G
11A	11B	11C	11D	11E	11F	11G

11 Rows of Lithology
1. Unconsolidated Sediment
2. Carbonate Sedimentary Rock
3. Mixed Sedimentary Rock
4. Non-Carbonate Sedimentary Rock
5. Evaporite
6. Pyroclastics
7. Metamorphic Rock
8. Acidic Volcanics
9. Acidic Plutonics
10. Non-Acidic Volcanics
11. Non-Acidic Plutonics

7 Columns of Land Cover
A. Bare Area
B. Sparse Vegetation
C. Grassland, Shrub, or Scrub
D. Mostly Cropland
E. Mostly Deciduous Forest
F. Mostly Needleleaf/Evergreen Forest
G. Swampy or Often Flooded

The ELU map contains 3,639 distinct ELU units, too complex for a simple legend. Instead, the inventory at left shows all possible combinations of the ELU input layers, and their color assignments. To interpret the diagram, first find the intersection of the temperature and moisture classes, and then select the appropriate column for landform, either plains, hills, or mountains. The submatrix of lithology (rows) against land cover (columns) is then presented, and the combination of lithology and land cover is then selected.

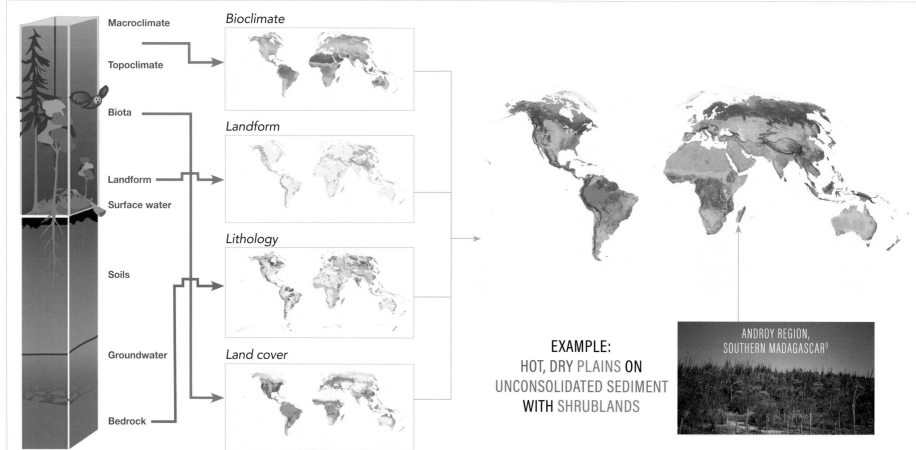

ELUs are mapped as distinct physical environments and associated vegetation. The layers—global bioclimates, global landforms, global geology, and global land cover—are first mapped individually and then combined in a GIS to create the specific ELU.

EXAMPLE:
HOT, DRY PLAINS ON UNCONSOLIDATED SEDIMENT WITH SHRUBLANDS

ANDROY REGION, SOUTHERN MADAGASCAR[9]

ELU TOOLS

Several query and visualization applications have been developed to allow easy access to the ELU data. These applications include a simple story map application (top figure), which introduces the users to the ELUs and allows them to pick any point on Earth and learn what ELU it is, including all of its building blocks (climate type, landform, geology, and land cover). This story map also provides a virtual tour to the 100 most ecophysiographically diverse areas ("ELU-speak" for a large variety of ELUs in one place), complete with stunning photos and rich descriptions of these unique landscapes.

Another application (middle figure) was built for the more scientifically inclined and data-proficient user. This sophisticated query browser goes beyond the simple "what is where" to a much richer, multivariable, simultaneous query. The user queries the map at any point and learns the ELU at that point, as expected, but the map also shows all the occurrences of that same ELU everywhere, basically the range of the ELU. Then, there is an additional set of four panels on the side, and each one shows the value of the four inputs (climate, landforms, geology, and land cover), not only at the selected point but also everywhere that particular input type is found. This app is an example of GIS enabling scientific understanding through improved visualization and query, which would not otherwise be possible.

Finally, to allow access and query to users of mobile devices, a field app featuring the ELU data (bottom figure) was also developed. Download the app, called Field Notes, onto your mobile device, go anywhere in the world, turn on the app, and it will characterize the landforms, climate region, geology, and land cover at your location.

Access all the online tools on this page at GISforScience.com.

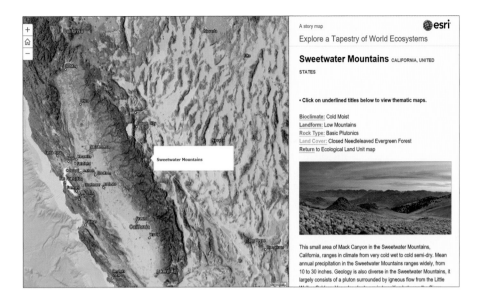

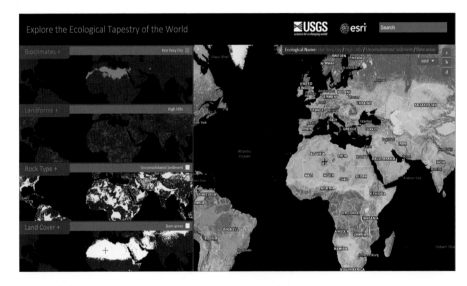

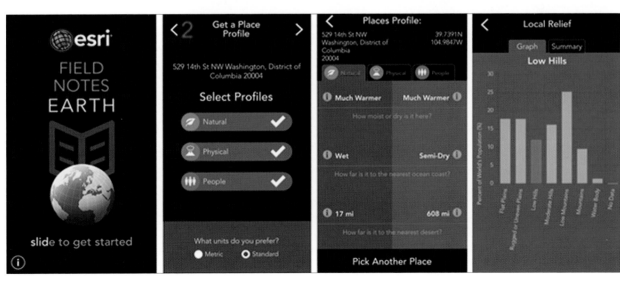

ECOLOGICAL MARINE UNITS

Having mapped and "apped" the terrestrial ecosystems of the planet, the GEM team then turned its attention to the global ocean, asking (1) What do we know about the ocean anyway? and (2) Is marine facet mapping a way forward? What is generally known about the ocean is often portrayed in a summary visual graphic such as the one shown on this page.[10] These kinds of graphics are found in many introductory oceanography textbooks. They suggest that there is vertical zonation in the water column (pelagic) of the ocean, with a relatively shallow sunlit zone (epipelagic) to about 200 m, then a twilight zone (mesopelagic) to about 1,000 m, and a dark zone (bathypelagic) deeper than that.

Considering that the nature of the ocean varies by depth, the mapping team recognized that it would be mapping facets in 3D. The ELUs on land were a 2D layer representing land facets on the surface of the earth. But mapping a set of ecological marine units would necessarily require mapping the ocean in 3D. The team thus set out to construct the first globally comprehensive, true 3D, data-derived map of distinct marine regions—not a 2D map of different regions on the ocean surface but a 3D map of different volumetric regions of the ocean.

Working with a map in 3D is where GIS enables science, as you'll discover again and again in this book. Not only does GIS permit visualization in 3D, a relatively recent breakthrough, but it is also increasingly used for spatial data interpolation and analysis. The team used a 57-year archive of data on the marine physical and chemical environment collected from 52 million points. Six variables were used as inputs to do the global mapping and analysis—temperature, salinity, dissolved oxygen, nitrate, phosphate, and silicate. The average value over the entire 57-year record for each of these variables at each of the 52 million points was calculated. Then an empty mesh, or grid, of points covering all the world's oceans was constructed in ArcGIS consisting of the 52 million locations. Think of the ocean mesh as a set of columnar stacks extending from the surface of the ocean to the seabed, at a regular spacing of approximately 27 km by 27 km.

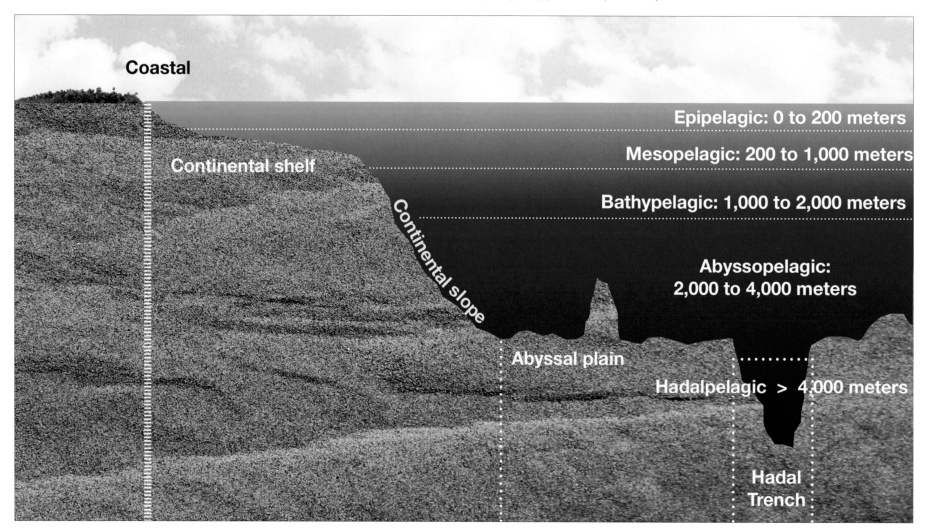

Coastal

Continental shelf

Continental slope

Epipelagic: 0 to 200 meters

Mesopelagic: 200 to 1,000 meters

Bathypelagic: 1,000 to 2,000 meters

Abyssopelagic: 2,000 to 4,000 meters

Abyssal plain

Hadalpelagic > 4,000 meters

Hadal Trench

EMU DATA ARCHITECTURE

Each individual stack (shown here) has several points at different depths, and each point contains the 57-year average values for the six variables. All the stacks together make up a global ocean wireframe of 52 million points. Once the data was partitioned, the team analyzed it using statistical clustering methods to tease out and separate volumetric regions that differed from each other based on their temperature, salinity, oxygen levels, and chemical constituents. In the end, 37 distinct EMUs, as volumetric regions, were mapped. Many were small in size and found in the upper ocean.

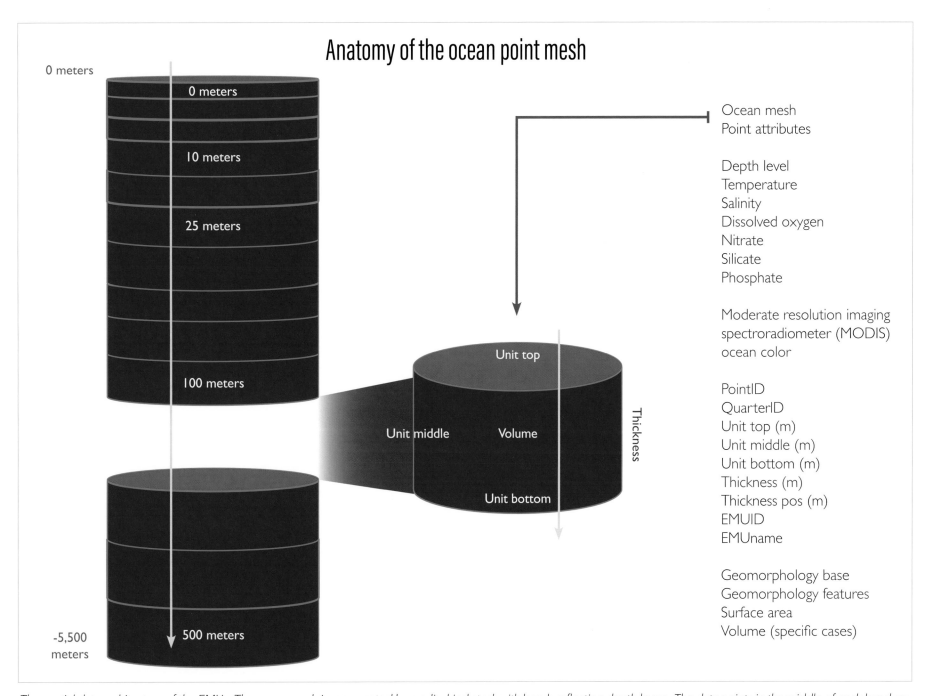

Anatomy of the ocean point mesh

0 meters

0 meters

10 meters

25 meters

100 meters

Unit top

Unit middle   Volume

Unit bottom

Thickness

-5,500 meters

500 meters

Ocean mesh
Point attributes

Depth level
Temperature
Salinity
Dissolved oxygen
Nitrate
Silicate
Phosphate

Moderate resolution imaging spectroradiometer (MODIS) ocean color

PointID
QuarterID
Unit top (m)
Unit middle (m)
Unit bottom (m)
Thickness (m)
Thickness pos (m)
EMUID
EMUname

Geomorphology base
Geomorphology features
Surface area
Volume (specific cases)

The spatial data architecture of the EMUs. The ocean mesh is represented by a cylindrical stack with bands reflecting depth layers. The data points in the middle of each band are attributed with the variables listed.

EMUs ADVANCE UNDERSTANDING OF DEPTH ZONES

The EMUs are a set of 3D ocean blobs that differ in their physical and chemical nature. The sheer number of attributes and depth levels requires the use of a GIS to map thin cross sections of data. Although difficult to see in their entirety in 3D, they can easily be seen by taking a 2D slice at any depth. For example, global EMU maps at the surface, at 1,000 m deep, and at 3,000 m deep, are depicted in the three depth maps. There is a considerable amount of data to explore. Imagine, for example, the ability to query and visualize the 52 million points, each with more than a dozen attributes, and then to query and visualize the 37 EMUs that the points belong to as well. The team developed an innovative visualization technique for seeing the EMUs in a 3D-like representation. As seen in the map on the facing page, the EMUs can be displayed as colored bands on a cylinder, where different colors depict different EMUs.

The EMUs are the first true 3D characterization of the global ocean, and the work was recently published in Oceanography.[11]

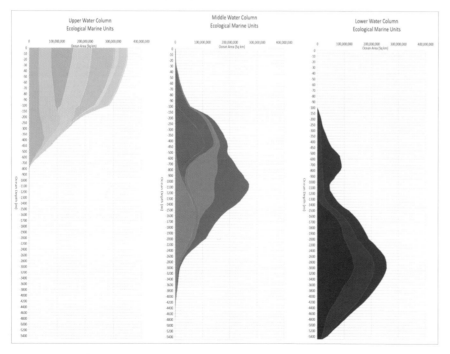

The EMUs got fewer and bigger with depth. Although the term blob mapping is hardly elegant given the ambition of the effort and the technological and analytical sophistication of the mapping approach, it nevertheless is quite useful in explaining the essence of EMUs.

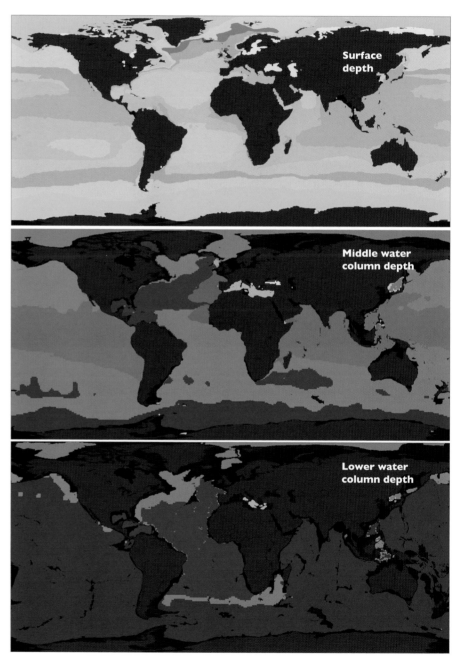

Global EMU maps at three depths. Top: EMUs at the sea surface. Middle: EMUs in the middle of the water column (depth = 1,000 m). Bottom: EMUs in the lower water column (depth = 3,000 m). Pinker colors indicate warmer EMUs; bluer colors indicate colder EMUs.

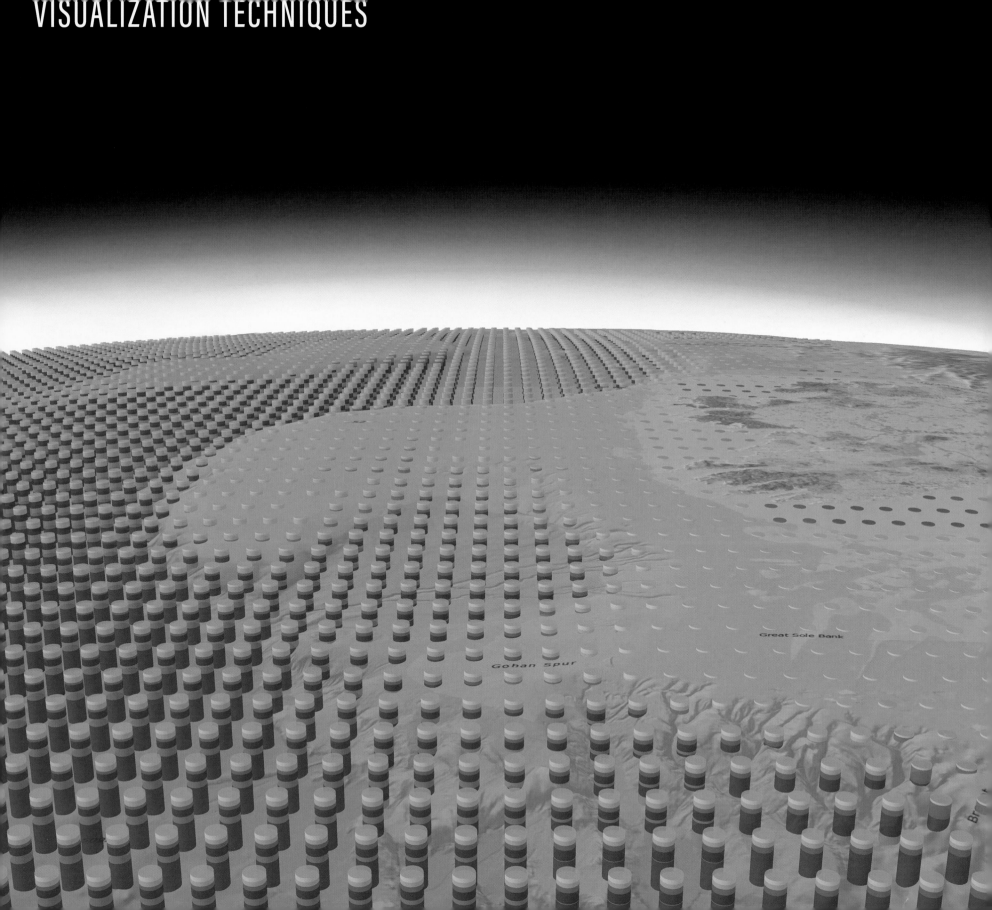

VISUALIZING EMUs

Although EMUs are in reality continuous data surfaces, they can instead be displayed as discrete bands on columns centered on the data points. This technique has the virtual effect of "cutting away" some of the data in between the columns so you can see down into the ocean and observe how the EMUs change both with depth and across space. This approach illustrates the spatial perspective at its finest and is yet another example of the blend of GIS technology and spatial science to enable the visualization and understanding of the global ocean in ways not heretofore possible. Using this visualization style, and adding ELUs into the mix to show terrestrial ecosystems, the team is now in a position to show EMUs and ELUs for any region on the planet. This map is an example of this mixed EMU and ELU mapping for the lands and ocean waters of Southern and Central California.

Cylinders, however, are not the only visualization objects that can be used to better comprehend 3D data. Even complex ocean properties such as currents, which have both a direction and a magnitude, can be visualized using innovative approaches developed by the team. The following image, for example, shows ocean current data in 3D (top) and ocean oxygen level in 3D (bottom) using arrows and varying diameter disks, respectively.

This graphic shows the innovative visualization approach taken to represent the EMUs as colored bands on cylinders centered on the data points. "Cutting away" the space between the cylinders permits looking down into the water column to see how EMUs change not only horizontally across space but also vertically through depth. Bands on cylinders represent different EMUs, and colors represent the temperature averages (pink, warm; blue, cold). Here, the EMUs in the ocean waters off Southern and Central California are shown, and on land the ELUs are shown.

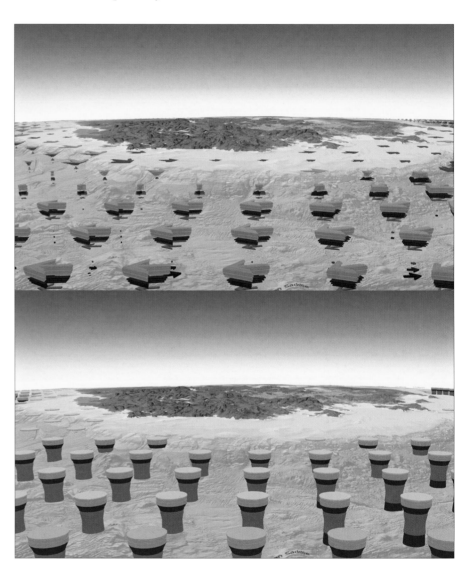

Different approaches to 3D visualization of two EMU attributes: ocean current velocity and direction (top) and oxygen depletion (bottom). For ocean currents, arrows indicate the direction of the current, and the arrows' size symbolizes the current's velocity (the larger the arrow, the faster the velocity). For oxygen depletion, the thickness of the band shows the concentration of dissolved oxygen (the thinner the band, the smaller the concentration of oxygen).

EMU EXPLORER

Having the new EMU data, maps, and visualization approaches available is a great step forward in the scientific understanding of the global ocean and how the ocean varies geographically. Like the ELUs before them, however, to make these resources more broadly available requires web-based and mobile apps that are easy to obtain and easy to use. These apps are available at GISforScience.com, this book's companion website. The team developed a web-based app for exploring the EMU data and immediately ported that technology to mobile devices, both iPhone® and Android®. The app is a vertical profiler and works on the pick-a-card, any card, notion brought forward from childhood memories of the quintessential card trick. In the case of our EMU app, the equivalent idea is pick-a-point, any point, anywhere on the surface of the ocean, and "the answer shall be revealed." The answer is, of course, the characteristics of that point, in terms of the temperature, salinity, and so on, and also the characteristics of the EMU that the point belongs to.

Beyond that, the app factors in every single point underneath the selected surface point on a line straight down to the sea floor. The app not only processes a tremendous amount of information (52 million points!) in real time but also reports back easy-to-understand summaries of interactive queries. The app is depicted on this page and shows the web-based vertical profiler (top) and the mobile-enabled version (bottom).

Ecological coastal units

Although the EMUs intersect the coast, the team recognized the need to map the coastal ecosystems independently from the EMUs. Coastal ecosystems are different from open ocean ecosystems because of the proximity of land and the heavy footprint of humans. Compared with EMUs, coastal ecosystems are typically smaller in scale, often linear in dimension, influenced by both terrestrial and aquatic processes, and home to some 40 percent of the global population.

A new, first-of-its-kind effort to map the coastal ecosystems of the planet at a high spatial resolution (30 m) is now under way, and a set of global ecological coastal units (ECUs) will be produced. A prerequisite to doing this work is having an accurate, high spatial resolution global shoreline vector (GSV) to separate land from water and to serve as the spatial and line work backbone of the effort. In terms of existing global shoreline representations, the ECU team did not find a suitable candidate to meet the following criteria: not proprietary (e.g., public domain), not crowdsourced (e.g., standardized and replicable), high spatial resolution (30 m or better), accurate, and current. The team is therefore developing a new 30 m GSV to use for ECU mapping. We will use the GSV to map the coastal land areas on one side of the line and coastal nearshore and offshore waters on the other side. We will then map the ECUs as unique coastal facets in terms of their environmental setting and biology.

Ecological freshwater units

As with the ELUs and the EMUs before them, the ECU effort will include publications, data, maps, and apps. When that work is concluded, the team will turn its attention to the remaining task of the original GEO global ecosystem mapping charge—freshwater ecosystems. We envision the development of global ecological freshwater units (EFUs) using the same kind of facet mapping wherein freshwater ecosystems (e.g., lakes, ponds, bogs, rivers, streams, and wetlands) are mapped as distinct combinations of physical environment setting and biological setting. This work will build on existing efforts to map global wetlands from space and ecologically classify global river reaches.

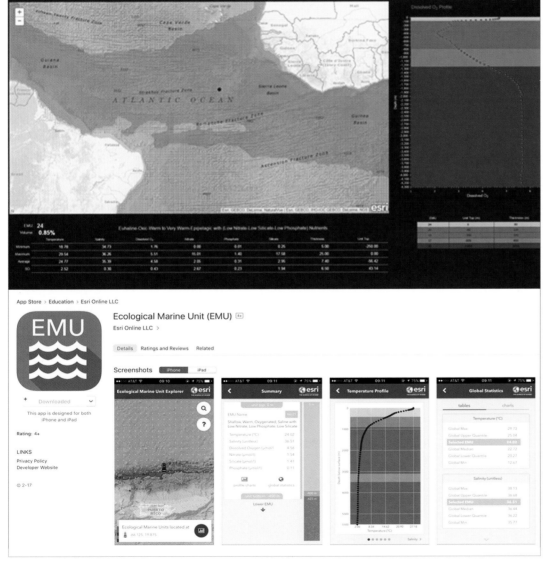

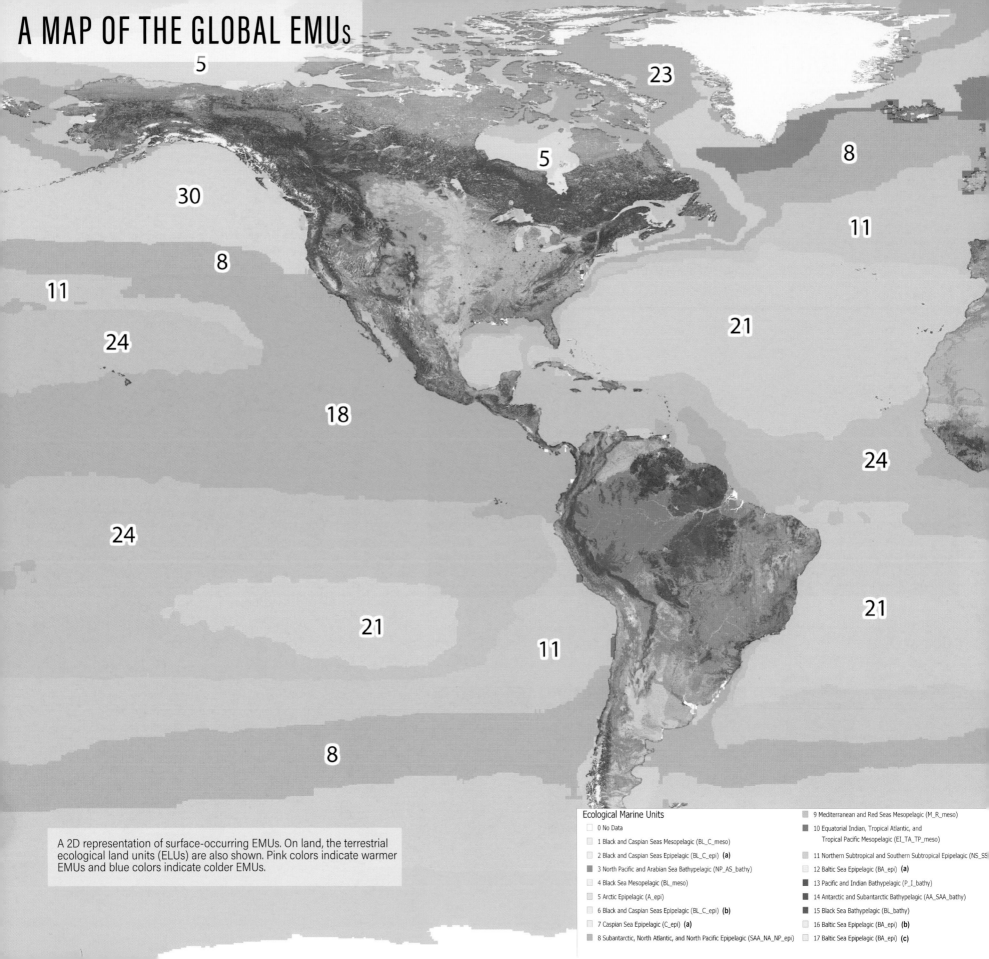

A MAP OF THE GLOBAL EMUs

A 2D representation of surface-occurring EMUs. On land, the terrestrial ecological land units (ELUs) are also shown. Pink colors indicate warmer EMUs and blue colors indicate colder EMUs.

Ecological Marine Units

- 0 No Data
- 1 Black and Caspian Seas Mesopelagic (BL_C_meso)
- 2 Black and Caspian Seas Epipelagic (BL_C_epi) **(a)**
- 3 North Pacific and Arabian Sea Bathypelagic (NP_AS_bathy)
- 4 Black Sea Mesopelagic (BL_meso)
- 5 Arctic Epipelagic (A_epi)
- 6 Black and Caspian Seas Epipelagic (BL_C_epi) **(b)**
- 7 Caspian Sea Epipelagic (C_epi) **(a)**
- 8 Subantarctic, North Atlantic, and North Pacific Epipelagic (SAA_NA_NP_epi)

- 9 Mediterranean and Red Seas Mesopelagic (M_R_meso)
- 10 Equatorial Indian, Tropical Atlantic, and Tropical Pacific Mesopelagic (EI_TA_TP_meso)
- 11 Northern Subtropical and Southern Subtropical Epipelagic (NS_SS
- 12 Baltic Sea Epipelagic (BA_epi) **(a)**
- 13 Pacific and Indian Bathypelagic (P_I_bathy)
- 14 Antarctic and Subantarctic Bathypelagic (AA_SAA_bathy)
- 15 Black Sea Bathypelagic (BL_bathy)
- 16 Baltic Sea Epipelagic (BA_epi) **(b)**
- 17 Baltic Sea Epipelagic (BA_epi) **(c)**

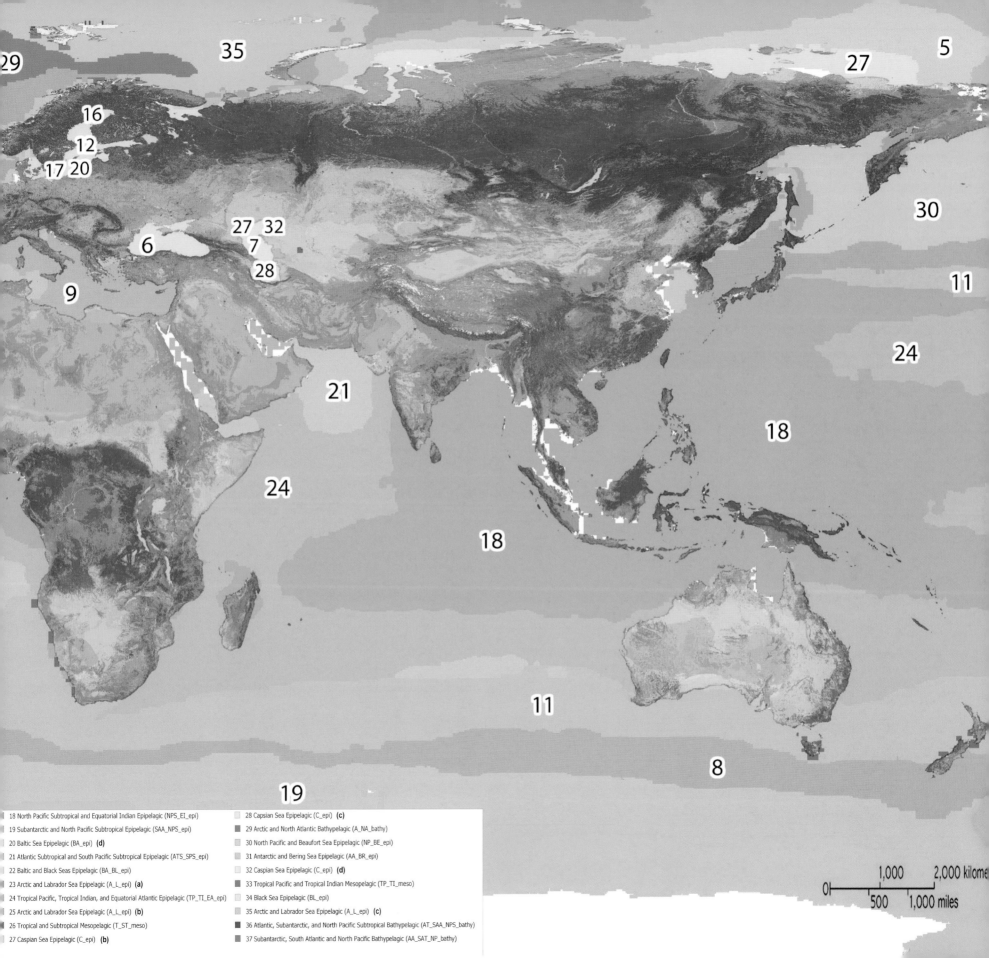

18 North Pacific Subtropical and Equatorial Indian Epipelagic (NPS_EI_epi)

19 Subantarctic and North Pacific Subtropical Epipelagic (SAA_NPS_epi)

20 Baltic Sea Epipelagic (BA_epi) **(d)**

21 Atlantic Subtropical and South Pacific Subtropical Epipelagic (ATS_SPS_epi)

22 Baltic and Black Seas Epipelagic (BA_BL_epi)

23 Arctic and Labrador Sea Epipelagic (A_L_epi) **(a)**

24 Tropical Pacific, Tropical Indian, and Equatorial Atlantic Epipelagic (TP_TI_EA_epi)

25 Arctic and Labrador Sea Epipelagic (A_L_epi) **(b)**

26 Tropical and Subtropical Mesopelagic (T_ST_meso)

27 Caspian Sea Epipelagic (C_epi) **(b)**

28 Capsian Sea Epipelagic (C_epi) **(c)**

29 Arctic and North Atlantic Bathypelagic (A_NA_bathy)

30 North Pacific and Beaufort Sea Epipelagic (NP_BE_epi)

31 Antarctic and Bering Sea Epipelagic (AA_BR_epi)

32 Caspian Sea Epipelagic (C_epi) **(d)**

33 Tropical Pacific and Tropical Indian Mesopelagic (TP_TI_meso)

34 Black Sea Epipelagic (BL_epi)

35 Arctic and Labrador Sea Epipelagic (A_L_epi) **(c)**

36 Atlantic, Subantarctic, and North Pacific Subtropical Bathypelagic (AT_SAA_NPS_bathy)

37 Subantarctic, South Atlantic and North Pacific Bathypelagic (AA_SAT_NP_bathy)

SPIN-OFFS AND APPLICATIONS

These resources represent rich new sources of information for understanding planetary ecosystems. As open data, they are available and being put to use in several interesting applications. For example, the ELU data includes landform type[12] as an attribute, including four classes of mountains (high mountains, scattered high mountains, low mountains, and scattered low mountains). Mountain classes have been extracted and used to build a Global Mountain Explorer (GME).[13] The GME shown here is another example of an app that brings ecosystem data to the masses in an easy-to-use, web-based (no local GIS software required!) platform, extending the reach of the data well beyond its intended initial audience of ecosystem scientists. Like the apps mentioned earlier, this app is available from this book's companion website at GISforScience.com. The GME app shown here allows anyone with an internet connection to query any location on the planet and find out if there are mountains there, and if so, what kind of mountains they are. The tool does that using the ELU data and two previous global mountain datasets, allowing the comparison of varying mountain definitions and different global mountain maps.

The GME built from landforms data associated with the ELU mapping effort. A screen capture of the interface is shown. The GME allows for the visualization and query of three global mountain characterizations.

The ELUs are being evaluated for use in support of the Intergovernmental Panel on Climate Change (IPCC) global emissions studies. IPCC modelers use a set of 20 generalized ecological zones as stratification areas within which to develop greenhouse gas inventories. The ELUs likely can be reconciled (crosswalked) with these ecological zones, bringing a finer spatial and thematic classification resolution to the set of stratification units currently in use.

The EMUs are also being evaluated for use in several different applications and value-added analyses.[14] Efforts are under way to assess the predictive potential of EMUs to delineate biogeographic regions (regions of differing biological [species] diversity), including the Atlas PROJECT[15] and a joint German-Russian exploration of the Northwest Pacific.[14]

NASA-funded research is under way to examine the relationship between satellite image-derived seascape maps and compositional data-derived EMUs.[14] Other research will address the integration of animal telemetry data with EMUs to better understand marine animal movements and behavior.

EMUs were used to predict suitable habitat conditions for global seagrass distribution under a scenario of increased ocean warming.[16] Results showed that a 1°C increase in ocean temperature may influence the temperature, salinity, oxygen, and nutrient regimes in a manner unfavorable for the persistence of seagrass, an important global carbon sink.

An important spin-off from the global EMU characterization has been the work to build a tool for the development of "localized" EMUs. Because the EMUs are global in nature and have a fairly coarse spatial resolution (~27 km by 27 km), they are often not suitable for local-scale, place-based assessments. The Esri team has therefore developed a tool and workflow for the development of the finest possible spatial resolution EMUs for a user-defined area. This web app is accessible from GISforScience.com. Finally, EMUs have been characterized as the type of "intuitive, interactive, dynamic online maps and visualizations"[17] needed to combat the increasingly recognized problem of ocean deoxygenation.

Conclusion

The need to understand the global distribution and condition of ecosystems is implicit in the UN SDGs, which call for the conservation of terrestrial, marine, and freshwater ecosystems. This work must be grounded in both ecological and spatial perspectives and will require advanced spatial science and technology to produce best-available characterizations of global ecosystem distributions. Applying The Science of Where to the Ecology of Earth has already resulted in the development of global ELUs (land ecosystems) and global EMUs (marine ecosystems). ECUs (coastal ecosystems) are in development, to be followed by EFUs (freshwater ecosystems). These resources, as publications, data, maps, and apps, are serving increasingly broader audiences and are advancing the science of ecosystem geography. They are also setting the stage for future assessments of ecosystem condition, change, and vulnerability, now that their distributions on land and in the sea are better understood. None of this progress would have been possible without the rapidly evolving technological advances in GIS for data manipulation, visualization, and analysis.

Earth is indeed a gem, and the more we see and understand its ecosystem facets, the more beautiful it becomes.

ENDNOTES

1. NCGE (National Council for Geographic Education). *Geography for Life: National Geography Standards,* 2nd ed. S.G. Heffron and R.M. Downs, eds., (Washington, DC: NCGE, 2014), 118.

2. United Nations Sustainable Development Goals: https://www.un.org/sustainabledevelopment/sustainable-development-goals/.

3. Graphic used with permission from Robert G. Bailey. Source: Bailey, R. *Ecoregion-based Design for Sustainability,* 222 (New York: Springer-Verlag, 2002).

4. J. E. M. Watson, K. R. Jones, R. A. Fuller, M. DiMarco, D. B. Segan, S. H. M. Butchart, J. R. Allan, E. McDonald-Madden, and O. Venter, "Persistent Disparities Between Recent Rates of Habitat Conversion and Protection and Implications for Future Global Conservation Targets," *Conservation Letters* 9, no. 6 (2016): 413–21, https://doi.org/10.1111/conl.12295.

5. Group on Earth Observations (GEO): http://www.earthobservations.org/index2.php.

6. US Geological Survey Science Strategy 2007–2009: https://pubs.usgs.gov/circ/2007/1309/pdf/C1309.pdf.

7. R. Sayre, J. Dangermond, C. Frye, R. Vaughan, P. Aniello, S. Breyer, D. Cribbs, D. Hopkins, R. Naumann, W. Derrenbacher, D. Wright, C. Brown, C. Convis, J. Smith, L. Benson, D. Van Sistine, H. Warner, J. Cress, J. Danielson, S. Hamann, T. Cecere, A. Reddy, D. Burton, A. Grosse, D. True, M. Metzger, J. Hartmann, N. Moosdorf, H. Durr, M. Paganini, P. DeFourny, O. Arino, S. Maynard, M. Anderson, and P. Comer, "A New Map of Global Ecological Land Units—An Ecophysiographic Stratification Approach" (Washington, DC: Association of American Geographers, 2014), 46.

8. R. Sayre, J. Dangermond, D. Wright, S. Breyer, K. A. Butler, K. Van Graafeiland, M. J. Costello, P. T. Harris, K. L. Goodin, M. T. Kavanaugh, N. Cressie, J. M. Guinotte, Z. Basher, P. N. Halpin, M. E. Monaco, P. Aniello, C. Frye, D. Stephens, P. Valentine, J. H. Smith, R. Smith, D. Paco Van Sistine, J. Cress, H. Warner, C. Brown, J. Steffenson, D. Cribbs, B. Van Esch, D. Hopkins, G. Noll, S. Kopp, and C. Convis, "A New Map of Ecological Marine Units—An Environmental Stratification Approach" (Washington, DC: Association of American Geographers, 2017), 35.

9. By Bernard DUPONT from FRANCE (Androy Scenery) [CC BY-SA 2.0 (https://creativecommons.org/licenses/by-sa/2.0)], via Wikimedia Commons.

10. Graphic used with permission from Hunter J. Sayre. Source: see Note 8.

11. R. Sayre, D. J. Wright, S. P. Breyer, K. A. Butler, K. Van Graafeiland, M. J. Costello, P. T. Harris, K. L. Goodin, J. M. Guinotte, Z. Basher, M. T. Kavanaugh, P. N. Halpin, M. E. Monaco, N. Cressie, P. Aniello, C. E. Frye, and D. Stephens, "A Three-dimensional Mapping of the Ocean Based on Environmental Data," *Oceanography* 30, no. 1 (2017): 90–103, https://doi.org/10.5670/oceanog.2017.116.

12. D. Karagulle, C. Frye, R. Sayre, S. Breyer, P. Aniello, R. Vaughan, and D. Wright, "Modeling Global Hammond Landform Regions from 250-m Elevation Data," *Transactions in GIS*, 22, no. 1 (2017), https://doi.org/10.1111/tgis.12265.

13. Global Mountain Explorer: https://rmgsc.cr.usgs.gov/gme/.

14. D. Wright, M. Kavanaugh, L. Henry, A. Brandt, H. Saeedi, N. Bednarsek, F. Muller-Karger, B. Ramiro-Sanchez, J. Roberts, T. Morato, K. Butler, K. Van Graafeiland, S. Breyer, and R. Sayre, "Use Cases of Ecological Marine Units for Improved Regional Ocean Observation Data Integration and Linkage Between End Users and Providers," *OceanObs'19 Proceedings* (2018), www.oceanobs19.net.

15. The ATLAS Project: https://medwavesblog.wordpress.com/atlasproject/.

16. N. Lanese, "Rising Ocean Temperatures Threaten Carbon-Storing Sea Grass." *Eos* (January 17, 2018). https://eos.org/articles/rising-ocean-temperatures-threaten-carbon-storing-sea-grass.

17. S. A. Earle, D. J. Wright, S. Joye, D. Laffoley, J. Baxter, C. Safina, and P. Elkis, "Ocean Deoxygenation: Time for Action," *Science* 359, no. 6383 (2018): 1475–76. https://www.doi.org/10.1126/science.aat0167.

Acknowledgments

The helpful reviews of Virginia Burkett and Shawn Komlos of the US Geological Survey were much appreciated.

Disclaimer

Any opinions expressed do not necessarily reflect the views or policies of the government of the United States. Any use of trade, product, or firm names does not imply endorsement by the US government.

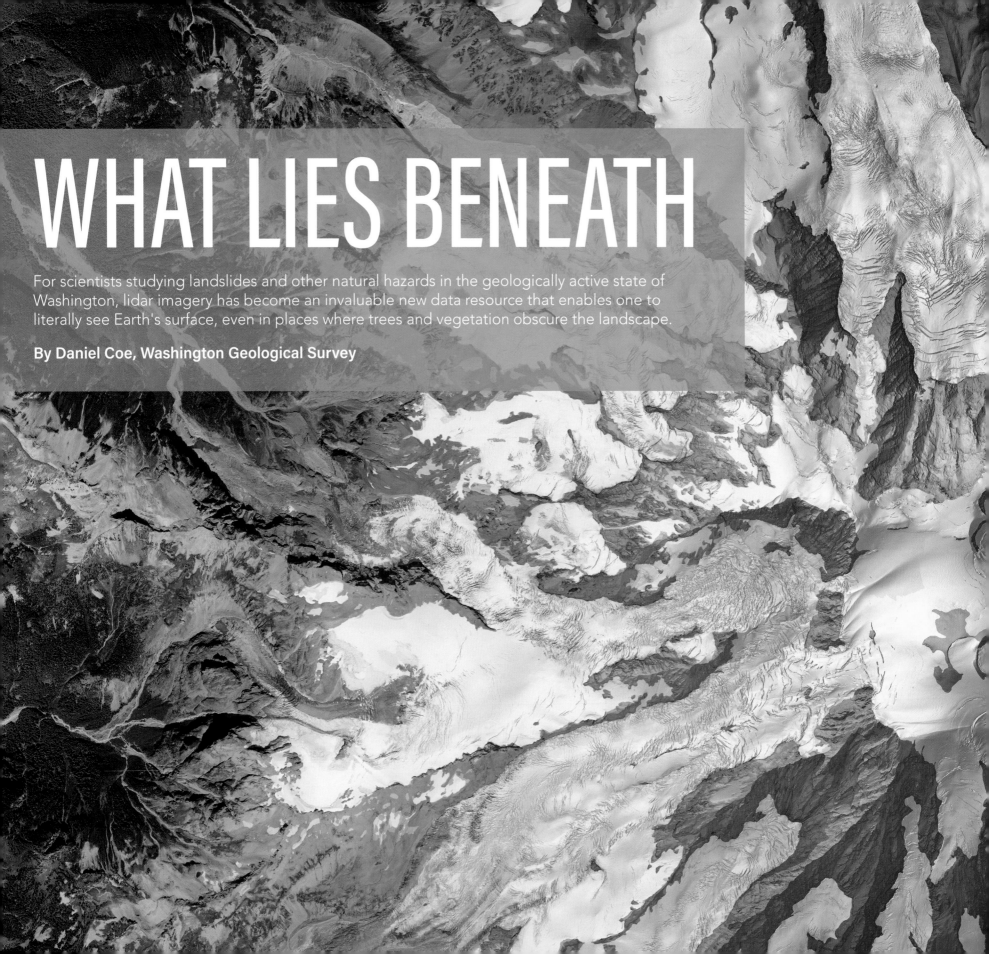

WHAT LIES BENEATH

For scientists studying landslides and other natural hazards in the geologically active state of Washington, lidar imagery has become an invaluable new data resource that enables one to literally see Earth's surface, even in places where trees and vegetation obscure the landscape.

By Daniel Coe, Washington Geological Survey

This lidar-shaded relief image delineates and emphasizes the glaciers and snowfields of Mount Rainier National Park.

WHAT IS LIDAR?

Light radar or light detection and ranging (lidar) technology uses light pulses to collect three-dimensional (3D) information. It is an invaluable, relatively new data source that enables geologists to see and study large areas of Earth's surface.

Aircraft often collect lidar data using a laser system pointed at the ground. The system measures the amount of time it takes for the laser light pulses to reach the ground and return. Billions of these rapidly collected measurements (points) can create extremely detailed 3D models of Earth's surface.

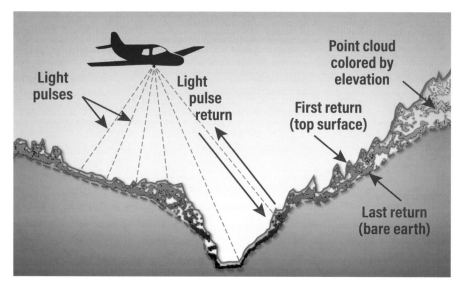

Airborne lidar collection, with the first return (top surface, solid jagged orange line) and last return (bare-earth, dashed smoother orange line) surfaces as well as a lidar point cloud, colored by elevation.

Oblique view of point cloud data from Beacon Rock in the Columbia Gorge. The colors in the image indicate elevations—lighter for higher elevations and darker for lower elevations.

Point cloud

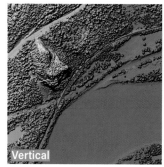

A collection of individual light pulse "returns" gathered by an aircraft. Vertical and oblique views of a lidar point cloud, showing all the light pulse returns. These colored elevation points include intensity data.

Top surface

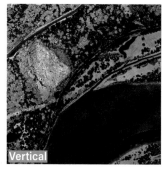

"First returns," including trees and structures interpolated into a continuous surface. The first returns can be digitally extracted to create a top surface model that includes the tops of trees and structures. This color elevation surface is rendered with shaded relief.

Bare earth

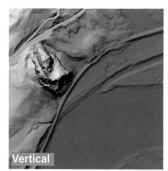

"Last returns" of the bare ground interpolated into a continuous surface. The last returns can be digitally extracted to create a bare-earth model of just the ground surface with all the trees and buildings removed. This color elevation surface is rendered with shaded relief.

BARE-EARTH LIDAR PRODUCTS

Geologists in Washington State use different bare-earth lidar products to map geology, landslides, and faults; to study volcanoes, glaciers, and rivers; and to model tsunami inundation. See the examples for a few of the different ways that bare-earth lidar can be displayed and analyzed using GIS tools.

In geology, lidar bare-earth models allow closer study of geomorphology, which is the study of the origin of the topography of Earth. Landslides, faults, floods, glaciers, and erosion leave their marks on the landscape, and although these marks can be hidden by dense vegetation, they can't hide from lidar. Visit GISforScience.com for links to online resources about more of the lidar applications developed by the Washington Geological Survey and to see more images of the state's intriguing landscape.

  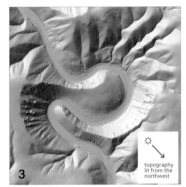  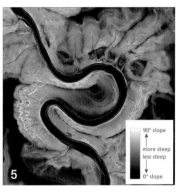

Five different lidar-derived analyses of Yakima Canyon, from left to right. (1) Digital elevation models (DEMS; also known as digital terrain models) show the elevation of an area using a color gradient. (2) Elevation contours show elevations and topographic relief using spaced contour lines. (3) A hillshade (also known as shaded relief) is a digitally illuminated view of the terrain, enhancing the 3D look of the landscape. (4) Aspect shows which direction slopes are facing. (5) Slope maps show the steepness of the land, from 0 to 90 degrees using a color gradient.

A blended image of the Cowlitz River with color natural light photography on the left merging into bare-earth lidar on the right.

LANDSLIDES

Landslides are among the most commonly occurring and devastating natural hazards in the state of Washington. In the past 30 years, the state has experienced numerous destructive landslides, including the 2014 State Route 530 "Oso" landslide that claimed 43 lives. Identification of potential landslide hazard areas is the first step toward reducing the impacts of landslides on property, infrastructure, and human life.

Before the widespread use of lidar, geologists used aerial photographs, topographic maps, and field surveys to identify landslides. These methods are problematic in much of Washington because the dense vegetation often obscures features and makes field checking difficult. Landslides often have characteristic topography. Slump blocks, hummocky topography, scarps, and sag ponds all can identify a landslide. With lidar, a geologist can search large tracts of land quickly and more accurately than using aerial photographs alone. To see the difference between lidar, photography, and older topographic data, see the images of the Devils Slide.

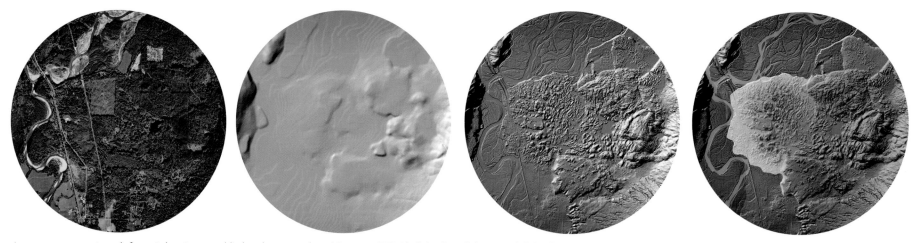

Imagery comparison left to right: A natural light photograph, a 10-meter DEM hillshade, a lidar gray hillshade, and a lidar colorized hillshade of Devils Slide (also known as the Van Zandt landslide) in Whatcom County, Washington. In the photograph, the landslide is virtually invisible. The 10-meter hillshade begins to suggest the landslide morphology, but in the lidar images, the landslide is obvious.

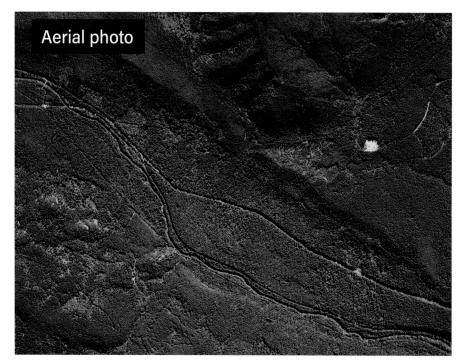

Aerial photo

Aerial photo from the Cedar River watershed; landslides are hidden underneath a green blanket of trees.

Bare-earth lidar

The bare-earth lidar image reveals multiple landslide headscarps and deposits along the river.

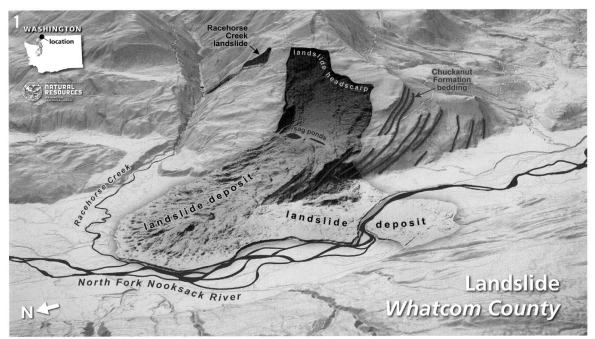

Landslide
Whatcom County

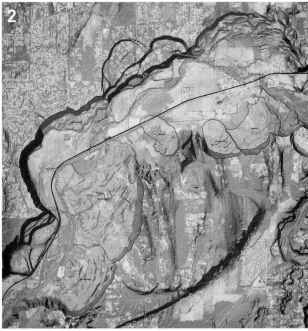

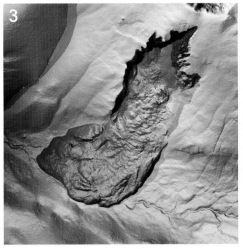

High-quality lidar maps of landslides reveal vivid and dramatic views of ancient (and recent) events. (1) Oblique view of a massive prehistoric landslide along the Nooksack River in northern Washington. (2) Landslide inventory mapping in Pierce County. (3) A lidar bare-earth image of a landslide in Mount Rainier National Park. (4) Oblique lidar image of large landslides above the White River in Pierce County.

Landslide inventory workflow

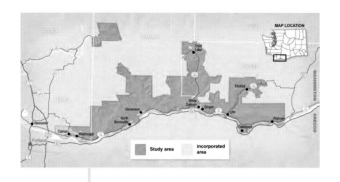

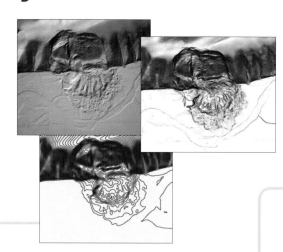

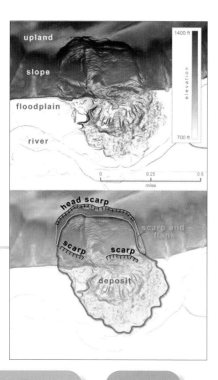

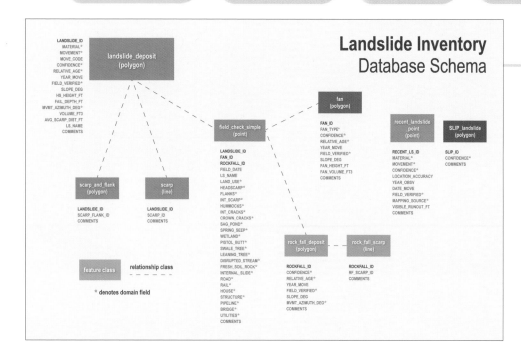

Determine study area

Based on proximity to population, highway corridors, waterways, and high-quality lidar availability.

Process lidar derivatives from digital elevation model

Hillshades, contour lines, slope maps

Interpret and delineate landslide landforms

Deposits, scarps, flanks

Assign attributes

Material, movement type and code, confidence, age, year moved, field verification, slope degree, headscarp height, failure depth, movement azimuth, and average scarp distance.

Field verification

>10 percent of mapped landslides field checked

Priority to populated areas and infrastructure

Licensed geologist review

Landslide Inventory
Database Schema

landslide_deposit (polygon)

LANDSLIDE_ID
MATERIAL*
MOVEMENT*
MOVE_CODE
CONFIDENCE*
RELATIVE_AGE*
YEAR_MOVE
FIELD_VERIFIED*
SLOPE_DEG
HS_HEIGHT_FT
FAIL_DEPTH_FT
MVMT_AZIMUTH_DEG*
VOLUME_FT3
AVG_SCARP_DIST_FT
LS_NAME
COMMENTS

fan (polygon)

field_check_simple (point)

LANDSLIDE_ID
FAN_ID
ROCKFALL_ID
FIELD_DATE
LS_NAME
LAND_USE*
HEADSCARP*
FLANKS*
INT_SCARP*
HUMMOCKS*
INT_CRACKS*
CROWN_CRACKS*
SAG_POND*
SPRING_SEEP*
WETLAND*
PISTOL_BUTT*
SWALE_TREE*
LEANING_TREE*
DISRUPTED_STREAM*
FRESH_SOIL_ROCK*
INTERNAL_SLIDE*
ROAD*
RAIL*
HOUSE*
STRUCTURE*
PIPELINE*
BRIDGE*
UTILITIES*
COMMENTS

fan
FAN_ID
FAN_TYPE*
CONFIDENCE*
RELATIVE_AGE*
YEAR_MOVE
FIELD_VERIFIED*
SLOPE_DEG
FAN_HEIGHT_FT
FAN_VOLUME_FT3
COMMENTS

recent_landslide_point (point)

RECENT_LS_ID
MATERIAL*
MOVEMENT*
CONFIDENCE*
LOCATION_ACCURACY
YEAR_OBSV
DATE_MOVE
FIELD_VERIFIED*
MAPPING_SOURCE*
VISIBLE_RUNOUT_FT
COMMENTS

SLIP_landslide (polygon)

SLIP_ID
CONFIDENCE*
COMMENTS

scarp_and_flank (polygon)

LANDSLIDE_ID
SCARP_FLANK_ID
COMMENTS

scarp (line)

LANDSLIDE_ID
SCARP_ID
COMMENTS

rock_fall_deposit (polygon)

ROCKFALL_ID
CONFIDENCE*
RELATIVE_AGE*
YEAR_MOVE
FIELD_VERIFIED*
SLOPE_DEG
MVMT_AZIMUTH_DEG*
COMMENTS

rock_fall_scarp (line)

ROCKFALL_ID
RF_SCARP_ID
COMMENTS

feature class   relationship class

* denotes domain field

Landslide susceptibility workflow

Deep

Shallow

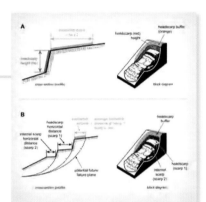

Factor of safety (FS) calculation using the infinite slope equation

Where c' is the effective cohesion, ϕ' is the effective angle of internal friction, α is the slope angle, γ is the unit weight (soil), γ_w is the unit weight of water, t is the slope-normal thickness of the potential slide block, and m is the proportion of slope thickness that is saturated.

$$FS = \frac{c'}{\gamma t \sin \alpha} + \frac{\tan \phi'}{\tan \alpha} - \frac{m \gamma_w \tan \phi'}{\gamma \tan \alpha}$$

high —FS less than or equal to 1.25

moderate —FS between 1.25 and 1.5

Generate high-susceptibility raster

Join deep landslide deposits with headscarp and flanks.

Calculate variable horizontal buffer.

Refine data

Use focal statistics, majority filter, and boundary clean tools.

Remove isolated areas that are < 45 square feet.

Buffer

Apply 30-foot buffer to all areas with FS < 1.5.

Generate moderate-susceptibility raster

Repeat variable horizontal buffer on high susceptibility zone.

Create generalized engineering geology map.

Combine landslide susceptibility variables: (1) geologic units, (2) contacts, (3) slope angles, and (4) preferred direction of movement.

Refine data

Remove isolated areas that are < 45 square feet.

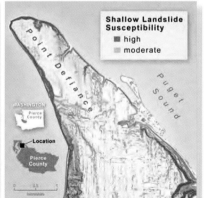

For more details on landslide inventory and susceptibility workflows, see protocols listed in the "Endnotes" section at the end of the chapter.

CASE STUDY: COLUMBIA GORGE LANDSLIDE INVENTORY

Landslides are common in the Columbia Gorge along the Washington–Oregon border because of the geologic setting, high relief, steep slopes, and abundant precipitation. In 2018, the Washington Geological Survey (WGS) published the results of a lidar-based landslide mapping project within a 900-square-mile area on the Washington side of the Columbia Gorge. The WGS mapped existing landslides and modeled where landslides might occur in the future. The new inventory contains 2,163 landslides covering approximately 16.5 percent of the study area. This updated landslide inventory and susceptibility analysis will increase awareness of landslide hazards in the Columbia Gorge and assist planners, emergency managers, public works departments, and those who live and work where landslides could impact their daily lives.

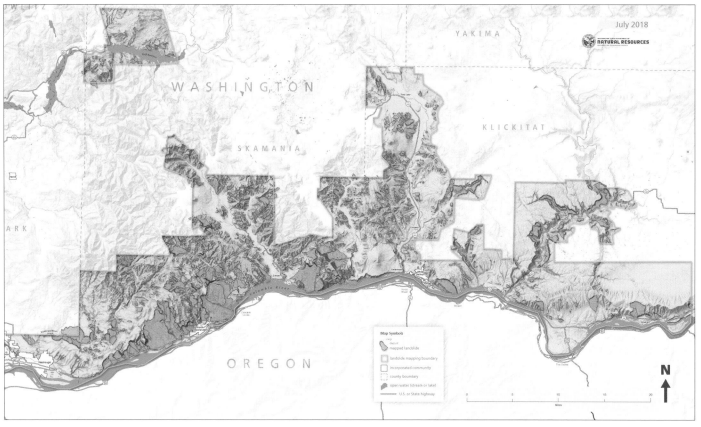

Map showing all the mapped landslides in the Columbia Gorge project area.

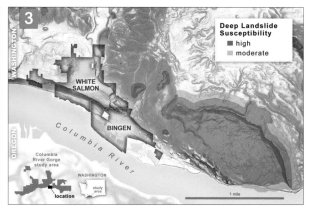

Below, WGS geologists field-checking a large landslide in the Columbia Gorge.

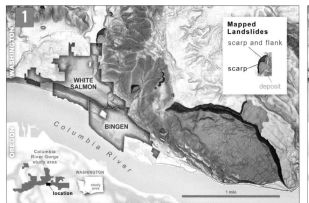

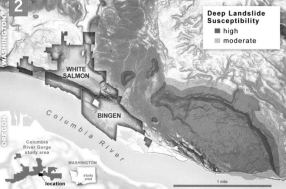

Detailed views of the (1) mapped landslides, (2) shallow landslide susceptibility, and (3) deep landslide susceptibility near White Salmon and Bingen, Washington.

VOLCANOES AND GLACIERS

Washington State has five major active volcanoes—Mount Baker, Glacier Peak, Mount Rainier, Mount Adams, and Mount St. Helens. All five are glaciated, and each has erupted in the past 250 years, except for Mount Adams. Many hazards come with living near volcanoes. Volcanic eruptions can send ash and volcanic debris into the air. The 1980 eruption of Mount St. Helens covered much of eastern Washington in ashfall that emitted from the mountain. Lava can also erupt and flow downhill, destroying everything in its path.

Heat from the volcano can melt glacial ice and snow and cause dangerous lahars. A lahar is a violent type of debris flow composed of a mixture of pyroclastic material, rocky debris, and water. The material flows down from a volcano, typically along a river valley. Lahars are extremely destructive: they can flow tens of meters per second. Lidar can be used to map areas of high lahar hazards. Accurate models of topography downstream from volcanoes can be used to predict lahar travel paths, pinpoint danger zones, and identify safe evacuation routes.
.

Photograph (left) versus bare-earth lidar image (right) of Mount Rainier. The lidar image emphasizes the glaciers and snowfields; in the photograph the snow and ice extents are more difficult to discern.

This composite lidar and photo comparison showing the USGS lahar hazard zone for the heavily populated Orting Valley, with Mount Rainier looming in the distance.

MONITORING MOUNT ST. HELENS

Washington's most notorious geologic event in recent history was undoubtedly the massive eruption of Mount St. Helens in 1980—the deadliest and most economically destructive volcanic event in US history. In geologic terms, the volcano is still extremely active, with a new dome forming in the crater. Repeated lidar surveys over time have been used to monitor volcanic activity within the volcano. While no new eruption is imminent, Mount St. Helens continues to change fairly rapidly, as seen in the two views taken seven years apart. This evidence allows geologists to accurately monitor and document dome building and the formation of new glaciers at the summit since 2002.

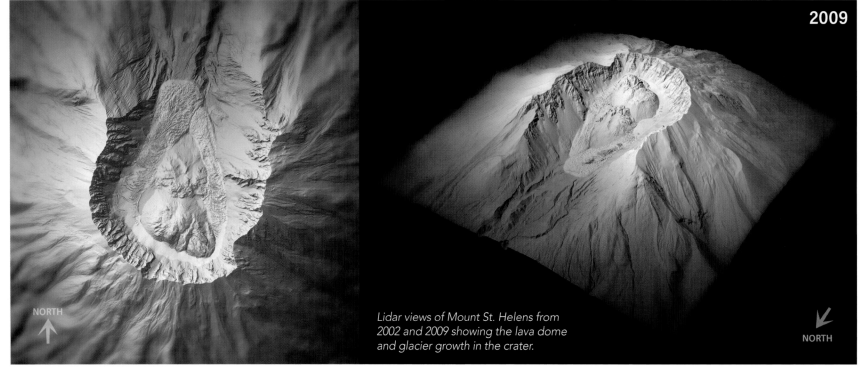

Lidar views of Mount St. Helens from 2002 and 2009 showing the lava dome and glacier growth in the crater.

Washington has several large glaciated peaks and mountain ranges. Lidar is used to monitor the growth or, more typically of late, decline of the glaciers in these locations. Repeated lidar flyovers over the same area document the effects of climate change and geologic activity on the snow and ice fields. It's now possible to discern the glacial history from the last ice age by identifying features such as moraines and outwash channels that appear in the wake of receding glaciers.

A lidar point-cloud image of Mount Baker in the North Cascades where glaciers have been receding continuously since the 1980s.

A bare-earth lidar color hillshade of the Carbon Glacier on Mount Rainier.

View of Mount Rainier, the most glaciated mountain in the lower 48 states, from White Pass.

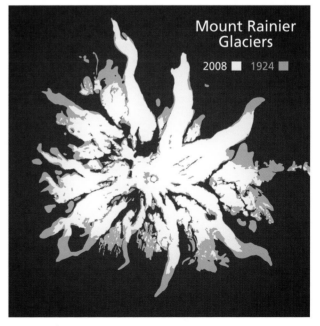

Glacier change on Mount Rainier between 1924 and 2008.

TSUNAMIS

Tsunamis constantly threaten the coastal regions of Washington. Lidar, in combination with bathymetry (topography beneath the water surface) data, allows models to more accurately predict where a tsunami could inundate an area. This modeling can give communities the best information for emergency planning and management.

In this image, modeled tsunami inundation covers much of the low-lying coastal landscape. This information can be used to plan evacuation routes.

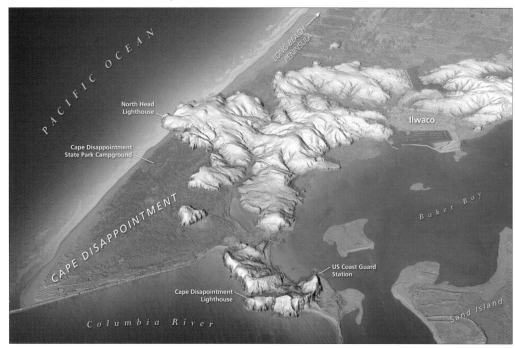

In this bare-earth lidar image, the contrast between the uplands and lowlands of Cape Disappointment is clearly visible.

A tsunami evacuation map of Cape Disappointment and the Long Beach Peninsula.

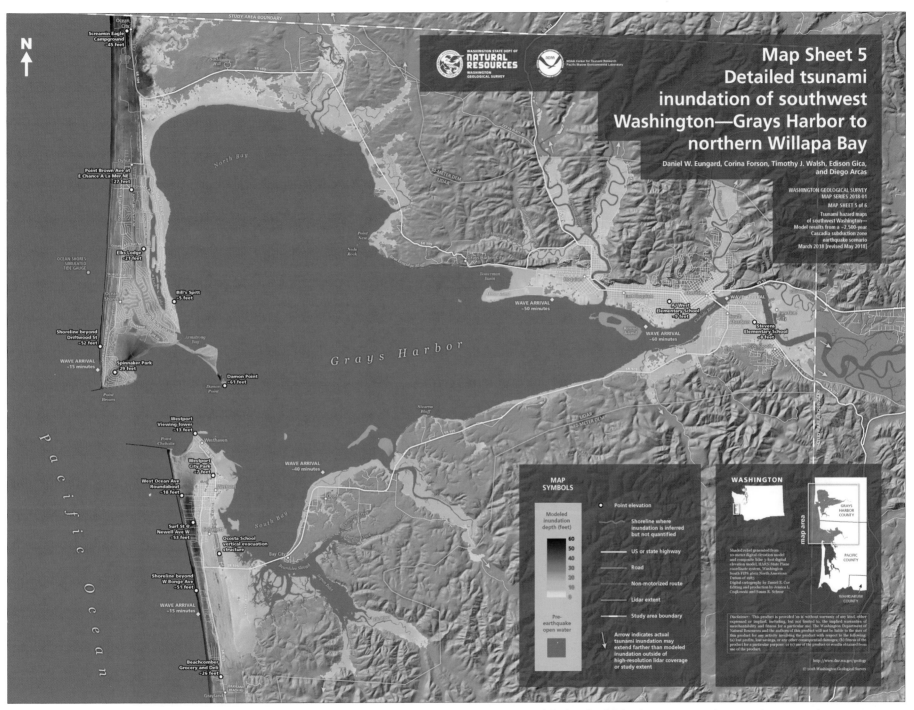

Map Sheet 5
Detailed tsunami inundation of southwest Washington—Grays Harbor to northern Willapa Bay

Daniel W. Eungard, Corina Forson, Timothy J. Walsh, Edison Gica, and Diego Arcas

WASHINGTON GEOLOGICAL SURVEY
MAP SERIES 2018-01
MAP SHEET 5 of 6
Tsunami hazard maps of southwest Washington—Model results from a ~2,500-year Cascadia subduction zone earthquake scenario
March 2018 [revised May 2018]

MAP SYMBOLS

- Point elevation

Modeled inundation depth (feet)
60
50
40
30
20
10
0

Shoreline where inundation is inferred but not quantified

US or state highway

Road

Non-motorized route

Lidar extent

Study area boundary

Pre-earthquake open water

Arrow indicates actual tsunami inundation may extend farther than modeled inundation outside of high-resolution lidar coverage or study extent

WASHINGTON

GRAYS HARBOR COUNTY

PACIFIC COUNTY

WAHKIAKUM COUNTY

Shaded relief generated from 10-meter digital elevation model and composite lidar 3-foot digital elevation model, HARN State Plane coordinate system, Washington South FIPS 4602 North American Datum of 1983
Digital cartography by Daniel E. Coe
Editing and production by Jessica L. Czajkowski and Susan R. Schnur

Disclaimer: This product is provided 'as is' without warranty of any kind, either expressed or implied, including, but not limited to, the implied warranties of merchantability and fitness for a particular use. The Washington Department of Natural Resources and the authors of this product will not be liable to the user of this product for any activity involving the product with respect to the following: (a) lost profits, lost savings, or any other consequential damages; (b) fitness of the product for a particular purpose; or (c) use of the product or results obtained from use of the product.

http://www.dnr.wa.gov/geology

© 2018 Washington Geological Survey

In 2018, the WGS, in partnership with the National Oceanic and Atmospheric Administration (NOAA), modeled and published tsunami inundation and current velocity maps for the southwestern coast of Washington. This project used bathymetric data combined with topographic lidar to model a simulated magnitude 9 earthquake event from the Cascadia subduction zone, a large, active plate boundary. The resulting maps show inundation depths ranging from 20 to 60 feet on the outer coast and generally less than 10 feet within Willapa Bay and Grays Harbor. The results of this study will be used for future evacuation and recovery planning for the communities in this region. The project was funded with a National Tsunami Hazard Mitigation Program grant.

RIVERS

Viewed through the unique lens of lidar imagery, the landscape of rivers and streams also reveals characteristics not obvious in traditional satellite imagery or aerial photography, making it useful for many hydrologic applications in the state. Floodplains can be mapped in detail to show where areas are at risk of flooding. Subtle river features, such as abandoned channels, ditches, terraces, and levees, also stand out in stark relief. Detailed floodplain mapping allows land managers and decision-makers to manage flood zones, preserve the natural functions of floodplains, and better craft emergency response procedures.

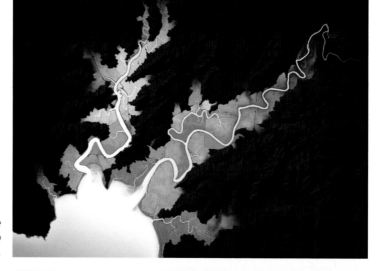

Levees are visible in this lidar image along the inter-tidal waterways of the Grays and Deep Rivers in Wahkiakum County, Washington.

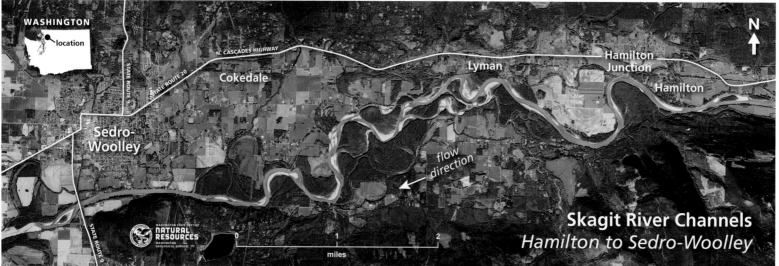

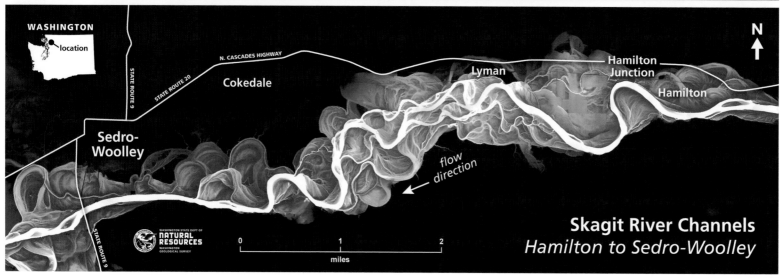

Lidar is especially useful for the visualization of large floodplains. This dual view of the Skagit River shows the current visible imagery (top) versus a lidar-derived relative elevation model (REM; bottom). As informative as it is beautiful, the REM imagery shows both current and former channels of the Skagit.

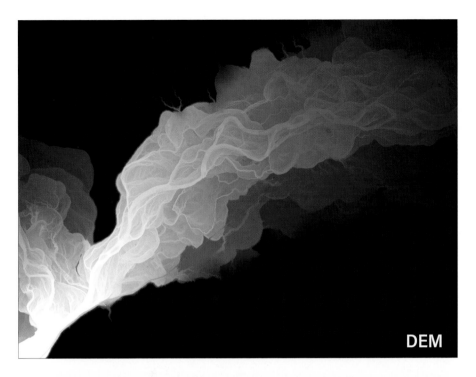

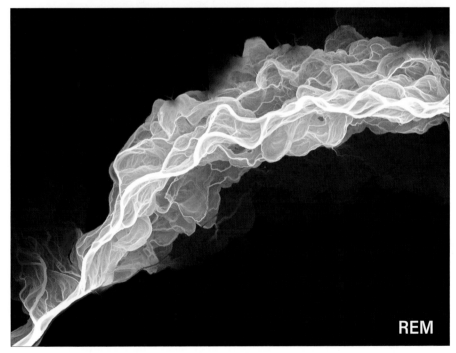

Sauk River comparison of a lidar DEM to a REM of the same area. In the DEM, standard elevations are displayed, while in the REM, all elevations are relative to the river surface, which is the base elevation throughout the image.

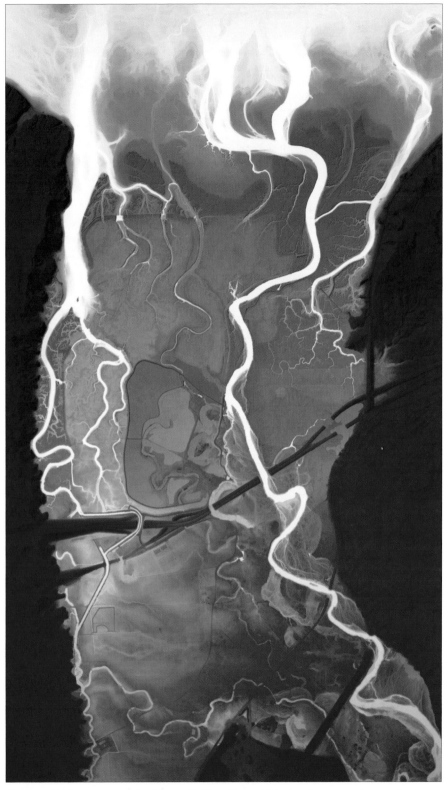

Bare-earth lidar image of the Nisqually River Delta. Lidar has been used to study the delta, where levees have been removed to reconnect tidal lands in an ongoing estuary restoration project.

SURFICIAL GEOLOGY

Lidar allows geologists to find subtle geologic features that might be otherwise missed. Features such as river terraces, drumlins, glacial outwash channels, and alluvial fans are readily apparent when high-quality lidar imagery is available. Landform analysis can be used to extrapolate field findings to correlate geologic units across different areas.

Glacial drumlins (fluted landforms) blanket the landscape in the bare-earth lidar image of the southern Hood Canal region.

The same flutes are barely visible in the aerial photograph.

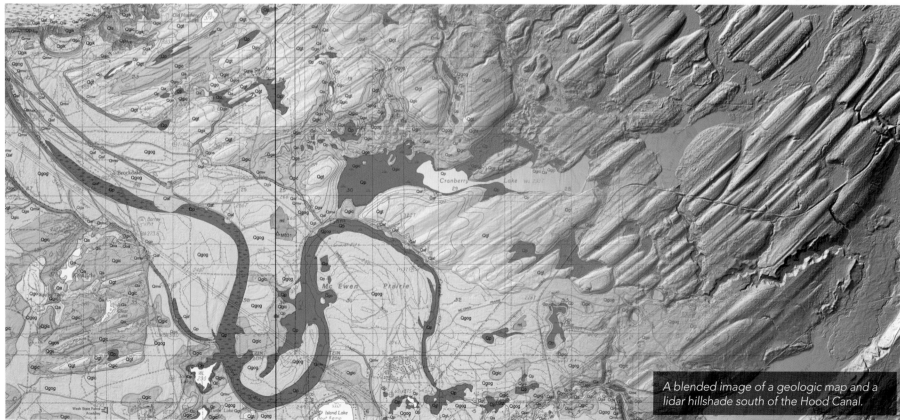

A blended image of a geologic map and a lidar hillshade south of the Hood Canal.

Glacial landforms of the Puget Lowland

During the last ice advance and retreat (in the latest Pleistocene), an extension of the Cordilleran ice sheet, called the Puget lobe, (covered the Puget Sound region (see location map below).

The colored area of the map (at right) represents the approximate maximum extent of the Puget Lobe during this time period. During glacial maximum, the location of modern-day Seattle (near the center of this map) was beneath 3,000 feet of ice.

All of the land and waterways in this region were shaped, at least in part, by the glacial ice of the Puget lobe. Many large-scale glacial landforms are preserved in the landscape today. In this lidar*-derived map, landforms such as drumlins, kettles, eskers, and glacial stream channels can be seen. Examples of these and other landscape features are enlarged at bottom right.

Most of Washington's population lives in this region—the glacial geology influences many aspects of daily life, including transportation, water supply systems, agriculture, and building regulations.

During the advance and retreat of the Puget lobe, drainages around the ice sheet were blocked, forming multiple proglacial lakes. The darker colors on this map indicate lower elevations, and show many of these valleys. The Stillaguamish, Snohomish, Snoqualmie, and Puyallup River valleys all once contained proglacial lakes. There are many remnants of these lakes left today, such as Lake Washington and Lake Sammamish, east of Seattle.

As the Puget lobe retreated, lake outflows, glacial meltwater, and glacial outburst flooding all contributed to dozens of channels that flowed southwest to the Chehalis River at the southwest corner of this map. Remnants of these channels can be seen along the eastern and southern edge of the colored area in the map. Present-day Lake Kapowsin and Ohop Lake both occupy one of these channels. Today, the Chehalis River flows through a wide valley that was largely sculpted by ice-age meltwater.

What is Lidar

The fine detail depicted on this map is the result of lidar. Lidar is a remote-sensing technique that uses light pulses to rapidly collect a very large quantity of elevation points across large areas. To learn more about lidar in Washington visit: www.dnr.wa.gov/lidar

NATURAL RESOURCES

www.dnr.wa.gov/geology   ©2017 Washington Geological Survey

1. **Drumlins** - Drumlins (or fluted ridges) are geologic features where movement of the ice sheet smooths glacial sediment into elongated teardrop shapes. Drumlins align in the direction of the ice flow and are evident across most of the Puget Lowland.

2. **Mima Mounds** - Intriguing features called Mima Mounds are found on Mima Prairie and in several outwash channels in the southern part of this map. Composed of organic-rich, sandy soil, Mima Mounds on this map are only found on the most recent glacial outwash deposits. The origin of the mounds has been debated for decades and a consensus on their formation has not yet been agreed upon.

3. **Kettles** - Glacial kettles are depressions that form when a retreating glacier leaves a bit of ice behind which then becomes buried by sediment shed from glacial streams. When the block of ice melts, the sediment collapses, forming a kettle. Kettles can be dry or filled with water, depending on their depth and the level of groundwater in the area.

4. **Eskers** - Eskers are snake-shaped landforms that are often found near the glacier's terminus. Eskers are formed when rivers that are underneath, on top of, or within a glacier, transport pebble- to cobble-sized gravel that is exposed once the glacier retreats. The resulting landform is a sinuous ridge of gravel that runs roughly parallel to the direction of ice flow.

5. **Outwash Channels** - Dozens of stream channels were created by glacial lake outflows, glacial meltwater streams, and glacial outburst floods. Today, many of the channels no longer transport water or have smaller streams occupying them than the streams that formed them.

6. **Fault scarps** - Sharp breaks in the fluted topography make it easy to identify geologically recent faults, such as this one called the Toe Jam Hill fault on Bainbridge Island west of Seattle, which is part of the Seattle fault zone.

This large-format map (shown here at 25 percent of intended scale) categorizes six different types of glacial landforms in the Puget Lowland that can be identified in lidar imagery: (1) drumlins, geologic features where movement of the ice sheet smooths glacial sediments into elongated teardrop shapes, (2) Mima Mounds, composed of organic-rich, sandy soil, (3) kettles, which form as depressions when a retreating glacier leaves behind a bit of ice that is buried in sediment, and then melts, (4) eskers, which are snake-shaped landforms, (5) outwash channels created by glacial lake outflows, meltwater streams, and outburst floods, and (6) fault scarps, which are easy to identify from the sharp breaks in the fluted topography.

BEDROCK GEOLOGY AND FAULTS

Lidar allows geologists to find structural and bedrock features such as folds and faults that indicate where Earth's crust has bent or broken. These features are often not visible on traditional topographic maps.

Because of plate tectonics, formations that make up Earth's crust are broken along faults. The past motion along faults is between centimeters and kilometers depending on the age and tectonic setting of the fault. Faults that break young deposits are considered potentially active or active. Active faults have the potential for rupture during earthquakes—many geologists study these features in detail to better estimate the frequency and magnitude of earthquakes along them. This information is essential in many activities such as road building and natural resource management.

Washington has dozens of active faults and fault zones. Some of these faults are in remote areas. Others, like the Seattle Fault and southern Whidbey Island fault zone, underlie major cities and pose a significant hazard.

Lidar imagery also helps scientists more accurately delineate contacts between geologic units in the field, sometimes revealing folds. Folds originally are flat surfaces, such as sedimentary strata, that are bent or curved because of tectonic movements. The traces of large folds are easily discernible from lidar.

Signs of young faults include sag ponds, offset streambeds, and linear scarps. Lidar gives geologists the ability to find these features no matter what the ground cover is like or if the feature is partly eroded.

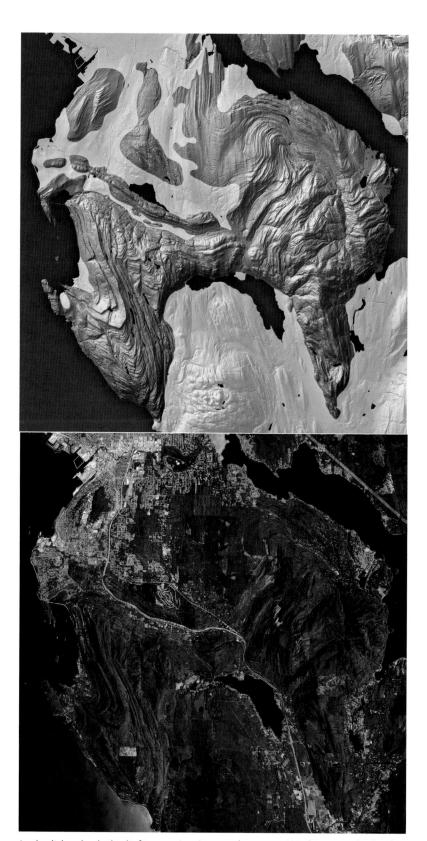

In the lidar-shaded relief image (top) in northwestern Washington, the bedding and folds of the Chuckanut Formation are starkly visible. In a photo of the same area (bottom), the full detail of the landscape is mostly lost underneath the thick forest and developed land.

ENDNOTES

This chapter was adapted from the story map: *The Bare Earth—How Lidar in Washington State Exposes Geology and Natural Hazards,* by the Washington Geological Survey.

Lidar data (2002–2018) from WGS, aerial ortho photography (2013–2017) from the National Agriculture Imagery Program (NAIP). All maps, lidar images, and graphics are from WGS. Ground-based photographs by Daniel E. Coe and Trevor A. Contreras, WGS. Additional photos by David K. Norman and Stephen L. Slaughter, WGS.

Burns, W. J., I. P. Madin, and K. A. Mickelson. 2012. "Protocol for Shallow-Landslide Susceptibility Mapping." *Special Paper* 45, p. 32, Oregon Department of Geology and Mineral Industries. http://www.oregongeology.org/pubs/sp/p-SP-45.htm

Burns, W. J., and K. A. Mickelson. 2016. "Protocol for Deep Landslide Susceptibility Mapping." *Special Paper* 48, p. 63, Oregon Department of Geology and Mineral Industries. http://www.oregongeology.org/pubs/sp/p-SP-48.htm.

Eungard, D. W., C. Forson, T. J. Walsh, E. Gica, and D. Arcas. 2018. *Tsunami Hazard Maps of Southwest Washington—Model Results from a ~2,500-year Cascadia Subduction Zone Earthquake Scenario.* WGS Map Series *2018-01*, originally published March 2018, six sheets, scale 1:48,000, 11 p. text. http://www.dnr.wa.gov/publications/ger_ms2018-01_tsunami_hazard_southwest_washington.zip.

Mickelson, K. A., K. E. Jacobacci, T. A. Contreras, W. Gallin, and S. L. Slaughter. 2018. *Landslide Inventory and Susceptibility of the Columbia Gorge in Clark, Skamania, and Klickitat Counties, Washington.* WGS Report of Investigation 40, p.11 text, with two accompanying Esri file geodatabases. https://fortress.wa.gov/dnr/geologydata/publications/ger_ri40_columbia_gorge_landslide_inventory.zip.

Mickelson, K. A., K. E. Jacobacci, T. A. Contreras, A. Biel, and S. L. Slaughter. 2017. "Landslide Inventory, Susceptibility, and Exposure Analysis of Pierce County, Washington." *WGS Report of Investigation* 39, p. 16 text, with two accompanying Esri file geodatabases and one Microsoft Excel file. https://fortress.wa.gov/dnr/geologydata/publications/ger_ri39_pierce_county_landslide_inventory.zip.

Polenz, Michael, T. A. Contreras, J. L. Czajkowski, G. L. Paulin, B. A. Miller, M. E. Martin, T. J. Walsh, R. L. Logan, R. J. Carson, C. N. Johnson, R. H. Skov, S. A. Mahan, and C. R. Cohan, 2010. "Supplement to Geologic Maps of the Lilliwaup, Skokomish Valley, and Union 7.5-minute Quadrangles, Mason County, Washington—Geologic Setting and Development Around the Great Bend of Hood Canal." *Washington Division of Geology and Earth Resources Open File Report*, p. 27.

Robinson, J. E., T. W. Sisson, and D. D. Swinney. 2010. Digital topographic map showing the extents of glacial ice and perennial snowfields at Mount Rainier, Washington, based on the LIDAR survey of September 2007 to October 2008. *U.S. Geological Survey Data Series* 549.

Slaughter, S. L., W. J. Burns, K. A. Mickelson, K. E. Jacobacci, A. Biel, and T. A. Contreras. 2017. "Protocol for landslide inventory mapping from lidar data in Washington State." *Washington Geological Survey Bulletin 82*, p. 27 text, with two accompanying Esri file geodatabases and one Microsoft Excel file. http://www.dnr.wa.gov/Publications/ger_b82_landslide_inventory_mapping_protocol.zip.

U.S. Geological Survey. 1924. *Mt. Rainier Quadrangle, Washington.* U.S. Geological Survey, 1 sheet, map scale 1:125,000.

https://www.dnr.wa.gov/geology.

The lidar data used to create the images in this chapter is available to the public through the Washington Lidar Portal. Here you will find links to downloadable digital surface models, digital terrain models, and point clouds. Other geologic data (including a veritable treasure trove of GIS-compatible data) that the WGS publishes can be accessed through the Washington Geologic Information Portal. Both of these reources are also linked from GISforScience.com.

THE ANATOMY OF SUPERVOLCANOES

Working in the shadows of some of the most remote volcanic regions on the planet, geologists use geostatistical analysis to reveal the space-time patterns of volcanic super-eruption in the Central Andes of South America.

By Melanie Brandmeier, Esri Germany

An approximately five-million-year-old, strongly altered volcano in Chaviña, southern Peru.

VOLCANISM

Volcanoes inspire many feelings in people: fear, passion, awe, as well as an innate curiosity about their origin. For ancient cultures, volcanoes represented the divine and inspired many legends, such as the tragic Aztec tale about Popocatépetl, a tribal leader who was transformed by the gods into a volcano to forever protect his dead love, Iztaccíhuatl. Popocatépetl near Mexico City is the most active volcano in Mexico. Modern science, however, tries to unravel the mechanisms involved in volcanic activity and related processes. This quest to understand is quite natural, as volcanic activity directly or indirectly affects our daily life in many ways: the formation of different types of ore deposits such as copper porphyries or epithermal gold deposits is linked to volcanic processes, fertile soils with high crop yields are found around volcanoes, geothermal energy is provided by volcanic systems, and even tourism is triggered by the attraction and beauty of volcanoes and related hot springs.

From time to time, we are reminded of the threat to civilization posed by volcanoes. In 1980, Mount St. Helens erupted and caused the death of 57 people in the state of Washington. In September 2018, a tsunami triggered by a volcanic eruption in Indonesia claimed over 2,000 lives. However, compared with the 1885 explosive eruption of Mount Tambora in Indonesia, costing the lives of some 93,000 people, this recent eruption was relatively minor. But even smaller eruptions can have far-reaching consequences: the 2010 eruption of Eyjafjallajökull, in Iceland, for example, inhibited flights over vast areas of Europe. Volcanic eruptions affect the environment in two primary ways: the direct impact by molten lava (magma) and pyroclastic flows, and the indirect effects caused by volcanic ash and gases being transported into the atmosphere, which can affect air quality, air travel, and even climatic patterns, depending on the magnitude and style of eruption. The most recent super-eruption on Earth occurred 26,500 years ago at Taupo, New Zealand. These rare supervolcanoes—and the role of GIS in uncovering previously unknown information about them—are the focus of this chapter.

Aerial view of the eruption of Mount St. Helens taken over Skamania County, Washington, on the morning of May 18, 1980.

From a fissure at Eyjafjallajökull, in Iceland, lava flows down toward the north, turning snow into steam.

The huge caldera (6 km in diameter and 1,100 m deep) formed when Tambora's estimated 4,000-meter-high peak was removed during the 1815 event—the most powerful volcanic eruption in historic times (VEI-7).

A BRIEF EXCURSION INTO VOLCANIC ERUPTION STYLES

In volcanology, we distinguish between different eruption types of volcanoes, where only one type of eruption might be observed in one particular volcano, or a whole sequence might occur during an eruptive series. In general, we distinguish three major types: magmatic eruptions, phreatomagmatic eruptions, and phreatic eruptions. Magmatic eruptions are the most common type and include Hawaiian, Strombolian, Vulcanian, Peléan, and Plinian subtypes that differ in their eruption explosivity and magma composition (as shown in the graphic). Besides shield volcanoes and stratovolcanoes, we need to briefly define calderas to understand Andean volcanism: generally, a caldera is a large basin-shaped crater at the top of a volcano. However, it is important to distinguish the large calderas forming by collapse over a large magma reservoir erupting pyroclastic flows (ignimbrites). Prominent examples are the Yellowstone Caldera, the La Pacana Caldera, and the Valles Caldera.

Supervolcanoes are defined by the volcanic explosivity index (VEI) that classifies the eruption of the volcano on the assumption of volume of the erupted products and the height of the ash columns. The index ranges from zero to eight; eruptions of six and more can cause a marked fall in global temperature (volcanic winter). A "supervolcano" is a volcanic center that has had an eruption of magnitude 8. It exceeds the scope of this chapter to go into more detail on eruption styles, magmatic rocks, and volcanic edifices. However, for people not familiar with geology, *Volcanism* by H.-U. Schmincke is probably one of the best reads on the topic.

Shield volcanoes   Increase of magma viscosity and explosivity   Stratovolcanoes

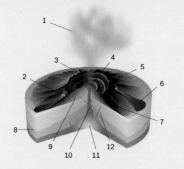
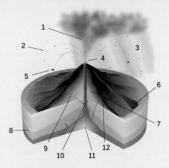
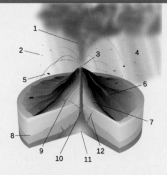
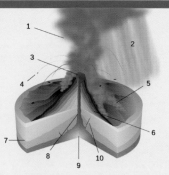
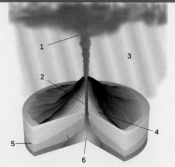

Hawaiian eruption:
1. Ash plume
2. Lava fountain
3. Crater
4. Lava lake
5. Fumaroles
6. Lava flow
7. Layers of lava and ash
8. Stratum
9. Sill
10. Magma conduit
11. Magma chamber
12. Dike

Strombolian eruption:
1. Water vapor cloud
2. Lapilli
3. Volcanic ash rain
4. Lava fountain
5. Volcanic bomb
6. Lava flow
7. Layers of lava and ash
8. Stratum
9. Dike
10. Magma conduit
11. Magma chamber
12. Sill

Vulcanian eruption:
1. Ash plume
2. Lapilli
3. Lava fountain
4. Volcanic ash rain
5. Volcanic bomb
6. Lava flow
7. Layers of lava and ash
8. Stratum
9. Sill
10. Magma conduit
11. Magma chamber
12. Dike

Peléan eruption:
1. Ash plume
2. Volcanic ash rain
3. Lava dome
4. Volcanic bomb
5. Pyroclastic flow
6. Layers of lava and ash
7. Stratum
8. Magma conduit
9. Magma chamber
10. Dike

Plinean eruption:
1. Ash plume
2. Magma conduit
3. Volcanic ash rain
4. Layers of lava and ash
5. Stratum
6. Magma chamber

Volcano diagrams © Sémhur / Wikimedia Commons / CC-BY-SA-3.0

Compare the relatively small magnitude of the Mount St. Helens eruption (red) to the Central Andean eruptions of Cerro Galán and La Pacana (Atana ignimbrite). Comparison of Eruption Volumes modified from USGS.

THE ANDES

The Central Andes is one of the most remote regions on Earth with landscapes dominated by salt flats and volcanoes. The region also contains among the largest ancient pyroclastic flows on Earth. Known as *ignimbrite*, the word comes from Latin *igni*, meaning fire, and *imbri* (rain). There are more than 185 active volcanoes in the Andes, giving the name "Andesite" to the magma type most prominent in Latin American volcanoes.

The Andes are an example of mountain building along an ocean-continent convergent plate boundary and relate to Mesozoic-Cenozoic subduction of oceanic lithosphere (mainly the Nazca plate) beneath the continental lithosphere of western South America. They constitute an approximately 8,000 km long and 250-to-750 km wide continuous topographic barrier in a north–south direction. This mountain chain can be geographically divided into the Northern, Central, and Southern Andes. The Central Andes consists (from west to east) of the Coastal Cordillera, the Longitudinal Valley, the Western and Eastern Cordilleras with elevations above 5,000 m, and the Altiplano plains with elevations of 3,800 m and 4,300 m.

With a crustal thickness of more than 70 km, the central volcanic zone (CVZ) of the Andes presently has the thickest crust in any subduction zone on Earth. The area has been an active volcanic arc since at least the Jurassic some 200 million years ago. However, shortening of the crust—the primary mechanism of crustal thickening—began no earlier than about 50 million years ago. It became more pronounced in the last 10 million years, leading to significant surface uplift and valley incision, as seen in the photo of Cotahuasi Canyon.

Many research groups and even SFBs (big cooperative German research projects), such as the one led by Onno Oncken, are dedicated to decipher processes of the formation of the Andes. Timing and spatial variations in Andean uplift and the formation of the Altiplano-Puna plateau are still a strongly debated issue in geodynamics. However, understanding processes and timing of uplift is crucially important to climatic and tectonic studies, as mountain ranges such as the Andes strongly affect climate patterns and may cause significant changes in global circulation. Thus, the Andes are a natural laboratory for the investigation of the interaction of mountain building, climate, and erosion.

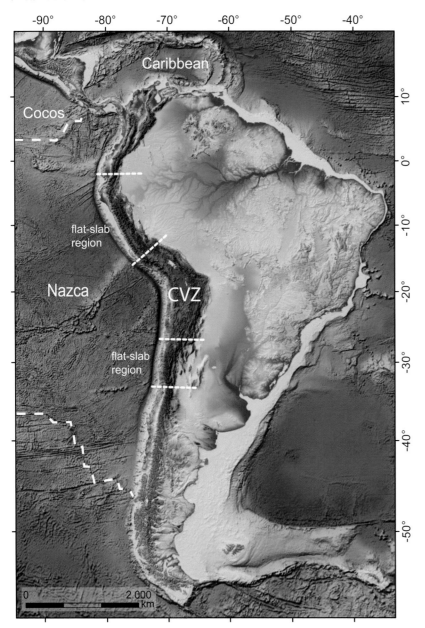

The world's longest continental mountain range, the Andean Mountains, runs almost the entire length of South America—north to south through seven South American countries: Venezuela, Colombia, Ecuador, Peru, Bolivia, Chile, and Argentina. The Central Andean Orocline between the northern and southern flat-slab segments represents the largest crust-forming volume that decreases toward the north and south central CVZ.

Cotahuasi Canyon (Peru). Alternating layers of ignimbrites and lava flows can be distinguished in the deeply incised valley. Volcanoes are sitting above the steep canyon flanks (background).

ANDEAN VOLCANISM

Volcanism in the CVZ during crustal thickening generally produced stratovolcanoes and cinder cones, but also ignimbrites that range in volume from a few cubic kilometers to several thousands of cubic kilometers, making the Central Andes one of the largest ignimbrite provinces on Earth. While steady-state volcanism of andesitic to dacitic lavas has generally dominated the history of volcanism in the Andes, massive ignimbrite eruptions also left their fingerprints on the landscape of the Central Andes.

Ignimbrites are pumice-dominated pyroclastic flow deposits with subordinate ash as shown in the photograph of the Altiplano. In the Andes, we distinguish between small-volume, valley-filling ignimbrites, and large-volume plateau-forming ignimbrites that result from so-called ignimbrite flare-ups sourced from large caldera complexes.

Ignimbrites that erupted in the CVZ during the Neogene—a geologic period that spans about 20.5 million years (20.5 Ma) from the end of the Paleogene Period to the beginning of the present Quaternary Period (2.58 Ma)—are variable in volume and composition. The largest ignimbrites are monotonous and crystal-rich, with volumes of individual flows reaching up to thousands of cubic kilometers.

Gigantic caldera structures such as the Vilama caldera, the La Pacana caldera, and the Cerro Galán complex, with diameters of tens of kilometers, are easily visible in high-resolution remote sensing data and can be mapped accordingly into GIS databases.

Ignimbrite sheets on the Altiplano, an area in west-central South America where the Andes are the widest. It is the second-most extensive area of high plateau on Earth after Tibet.

CERRO GALÁN AS AN EXAMPLE FOR LARGE CALDERAS

The Cerro Galán is a caldera in Catamarca Province, Argentina, one of the largest exposed calderas in the world and the youngest of the Altiplano-Puna-Volcanic Complex. It is well studied and described in detail by Francis et al. (1983) and others.

Its topographic caldera has dimensions of 38 by 26 km and is located at the intersection of the Archibarca-Galán lineament and the Diablillos-Galán Fault Zone and therefore a tectonically very active area. Due to the caldera's enormous size, you'd probably not notice driving within the caldera while crossing from one end to the other.

Galán was active between about 6.4 and 2 million years ago, when it generated a number of ignimbrites known as the Toconquis Group, which crop out especially west of the caldera. 2 Ma ± 0.02 years ago, the largest eruption of Cerro Galán was the source of the Galán ignimbrite, which covered the surroundings of the caldera. After this eruption, much smaller ignimbrite eruptions took place. The volume of all ignimbrites has been estimated to be about 1,200 cubic km.

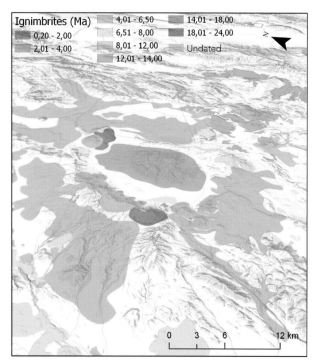

Ignimbrites (Ma)
0,20 - 2,00	4,01 - 6,50	14,01 - 18,00
2,01 - 4,00	6,51 - 8,00	18,01 - 24,00
	8,01 - 12,00	Undated
	12,01 - 14,00	

0   3   6   12 km

The map shows the Cerro Galán caldera and ignimbrite at present extent.

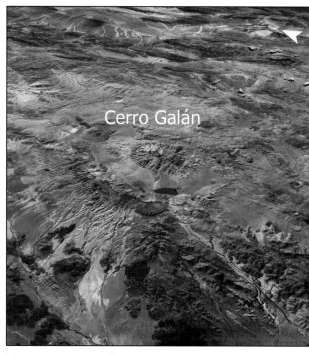

Cerro Galán

The same extent in the satellite image view.

The tranquility of the landscape of Cerro Galán is treacherous. Only two million years ago, some of the largest ignimbrite eruptions were sourced from this caldera.

WORKING IN PERU

The close relationship between deformation, uplift, magmatism, crustal growth, and ore formation makes the Latin American convergent plate margin an excellent setting to study the external and internal mechanism responsible for the evolution of the Andes and the Altiplano Puna Plateau, the second-highest plateau in the world. However, to investigate spatial and temporal patterns, a database containing different kinds of information layers is needed and was constructed over several years.

During two extensive field campaigns in 2011 and 2012, sampling was conducted in Southern Peru to close some knowledge gaps in this remote and difficult-to-access area. The goal of these campaigns was to collect ignimbrite samples for dating and geochemical analysis as well as to teach students about Andean geology and fieldwork. In total, more than 60 rock samples were collected, geotagged, and shipped to Germany to be analyzed at the University of Göttingen and dated at the University of Wisconsin (WiscAr Geochronology Labs).

Interestingly, massive ignimbrite deposits are not just geologic oddities: the remants of these ancient eruptions have been used as a source of building material called *sillar* for centuries. Quarried sillar is a whitish-pink rock used to build casonas, churches, and public buildings dating back to colonial days. Quarry workers continue to extract sillar for construction purposes; these days the demand primarily comes from new towns that are developing in the Arequipa region.

Data collected during the fieldwork helped to extend our knowledge of Andean ignimbrites from the well-studied Central Andes where mapping of ignimbrites is less difficult because of the arid climate compared with the more humid and less accessible areas of southern Peru.

Field excursion with students on the Altiplano.

Carved from the face of a massive ignimbrite at the Sillar Route near Arequipa in southern Peru, quarry workers have replicated the facade of the Church of the Company of Jesus, a landmark in Arequipa's historic city center, artistically carving it into the wall of the ravine.

THE ANDES IGNIMBRITE DATABASE

Researchers from inside and outside South America (see credits) have been studying and mapping ignimbrite formations in the Central Andes over time. The Andes Ignimbrite Database (AIDA) is the first attempt to compile all the known, reliable data into a central geospatial respository that would be accessible to the entire volcanic research community.

The ignimbrite flows were mapped using data collected during our fieldwork, published data, or traced from remote sensing data using ArcGIS imagery basemaps and ASTER data. (The Advanced Spaceborne Thermal Emission and Reflection Radiometer is an imaging instrument onboard Terra, the flagship satellite of NASA's Earth Observing System [EOS] launched in December 1999.) Information about age, thickness, and a brief description (where available) are included along with the exact location and shape of the flow. The areal extent and volume for each ignimbrite were then calculated (using an equal area projection) for previously mapped ignimbrites lacking this information.

The final GIS database includes 201 mapped ignimbrite sheets and caldera structures with information about age, area covered, estimated volume, and petrography. Furthermore, the locations of 1,672 ignimbrite samples with bulk-rock major and trace elements as well as isotopic data were included. The Andes Ignimbrite Database can be found linked at the companion website GISforScience.com. The availability of such datasets is highly important with respect to reproducibility of research as well as for further extending this work. By bringing together large amounts of data from literature and by collecting new data in the field, it is now possible to use machine-learning techniques—which require large amounts of data on an Andean scale—to find patterns in the geochemical and volumetric data and to link these patterns to the temporal evolution of the orogen.

The data is hosted in ArcGIS Online both as an app and as individual datasets that can be downloaded and integrated into desktop GIS software or other analytical environments and, importantly, onto mobile devices for reference during fieldwork.

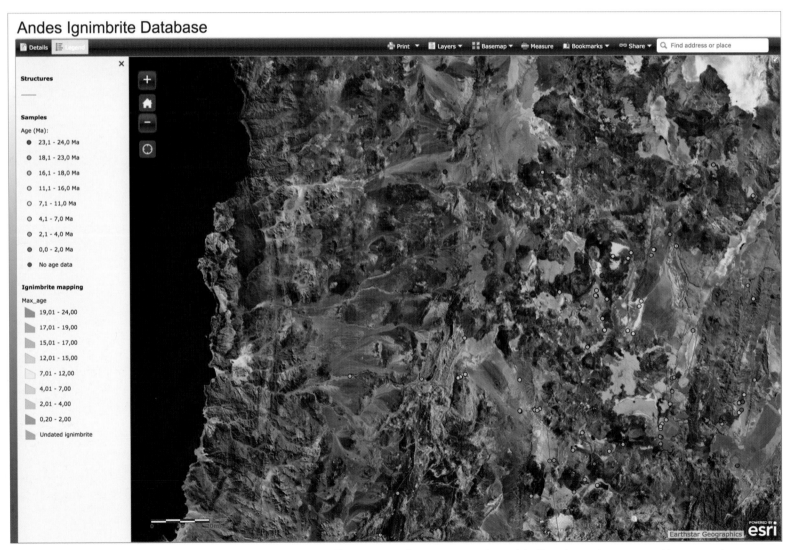

The Andes Ignimbrite Database contains about 200 mapped ignimbrites as well as 1,672 samples with bulk-rock geochemistry and/or age and is linked from GISforScience.com.

WORKFLOW

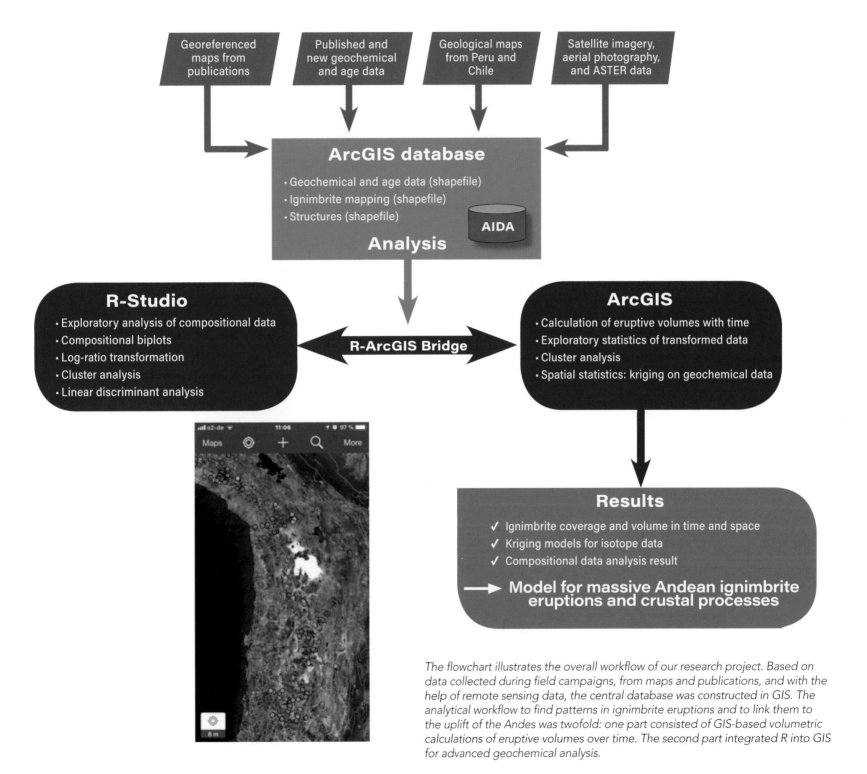

Georeferenced maps from publications

Published and new geochemical and age data

Geological maps from Peru and Chile

Satellite imagery, aerial photography, and ASTER data

ArcGIS database

- Geochemical and age data (shapefile)
- Ignimbrite mapping (shapefile)
- Structures (shapefile)

AIDA

Analysis

R-Studio

- Exploratory analysis of compositional data
- Compositional biplots
- Log-ratio transformation
- Cluster analysis
- Linear discriminant analysis

R-ArcGIS Bridge

ArcGIS

- Calculation of eruptive volumes with time
- Exploratory statistics of transformed data
- Cluster analysis
- Spatial statistics: kriging on geochemical data

Results

✓ Ignimbrite coverage and volume in time and space
✓ Kriging models for isotope data
✓ Compositional data analysis result

→ **Model for massive Andean ignimbrite eruptions and crustal processes**

The flowchart illustrates the overall workflow of our research project. Based on data collected during field campaigns, from maps and publications, and with the help of remote sensing data, the central database was constructed in GIS. The analytical workflow to find patterns in ignimbrite eruptions and to link them to the uplift of the Andes was twofold: one part consisted of GIS-based volumetric calculations of eruptive volumes over time. The second part integrated R into GIS for advanced geochemical analysis.

Ignimbrite mapping

Max_age

- 19,01 - 24,00
- 17,01 - 19,00
- 15,01 - 17,00
- 12,01 - 15,00
- 7,01 - 12,00
- 4,01 - 7,00
- 2,01 - 4,00
- 0,20 - 2,00
- Undated ignimbrite

Samples

Age (Ma):

- 23,1 - 24,0 Ma
- 18,1 - 23,0 Ma
- 16,1 - 18,0 Ma
- 11,1 - 16,0 Ma
- 7,1 - 11,0 Ma
- 4,1 - 7,0 Ma
- 2,1 - 4,0 Ma
- 0,0 - 2,0 Ma
- No age data

N

10 km

Cerro Galán

Zoom to the Cerro Galán Caldera (eastern part of the image) with several ignimbrites related to the caldera. In the western part of the image, some older ignimbrites stand out.

SPACE-TIME PATTERNS IN FLARE-UPS

The analytical workflow to decipher systematic patterns in super-eruptions is twofold (see Workflow, page 51): The first part is focused on machine-learning algorithms applied to log-ratio transformed compositional data to systematically analyze the GIS database on the scale of the entire Central Andean Orogen. Spatial analysis based on the ArcGIS platform is extended using the R-ArcGIS bridge that allows users to take advantage of the power available in R—a popular programming language for statistical analysis.

The second part of the analysis involves GIS-based calculations of eruptive volumes and the distribution of these eruptions in time and space. Both parts of the workflow are then brought together in a model that explains, together with studies by other researchers (see credits), eruptive patterns and causes of Andean super-eruptions.

Geochemical evidence—
the integration of R

Geochemical data has a distinct dataspace that does not allow many commonly used statistical methods without prior transformation. For very specific methods, the most advanced algorithms are implemented in R packages such as the package "compositions" used in this study.

Centered log-ratio transformation (clr) is a method to transform compositional data that allows analysis of relative magnitudes and variations of components (elements) rather than their absolute value. This enables the application of statistical methods based on the covariance or correlation matrix of vectors such as cluster analysis that are not otherwise possible due to the constraints of closed data. Cluster analysis is a method of unsupervised machine learning, which aims to define objects into groups ("clusters") so that the degree of association between two objects is maximal if they belong to the same group and minimal otherwise. It helps to discover structure in data without implying why it exists. Results of the cluster analysis based on rare earth elements (REEs) are shown on this page. Spatial dependencies, however, only become obvious when classifying and analyzing data according to volume and age. Such analysis reveals that ignimbrites show a significant change in composition over time and space.

A: Ternary diagram showing variations of clustering results for (REEs) as an example for compositional data analysis between La, Eu, and Yb (clr transformed and centered data); B: Compositional biplot (first and second Principal Component [PC]) C: REEs of respective clusters normalized to chondrite (McDonough and Sun 1995). Modified from Brandmeier and Wörner (2016).

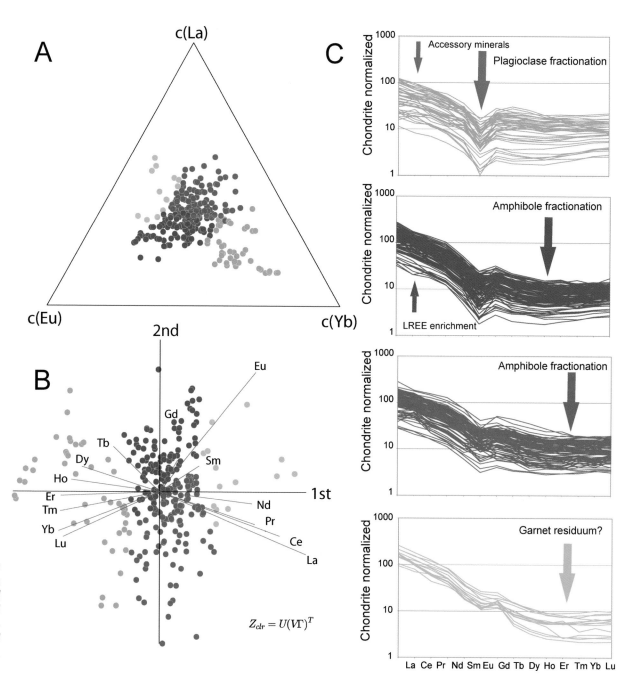

$$Z_{clr} = U(V\Gamma)^T$$

Volumetric variations by location

Surface area and volume were calculated in GIS for all ignimbrites in the database and separated into five north-to-south segments to analyze patterns of spatio-temporal variations in eruptive volumes. Even though such calculations remain imprecise because of a lack of sufficient data, several striking patterns are becoming obvious. Results are shown on the map of major ignimbrites and can be spatially related to the southward passage of the Juán Fernández Ridge over time. In 2009, Kay and Coira proposed the passage of the ridge to have caused the massive eruptions in the Puna because of mantle melting. This concept could now be extended northward toward southern Peru.

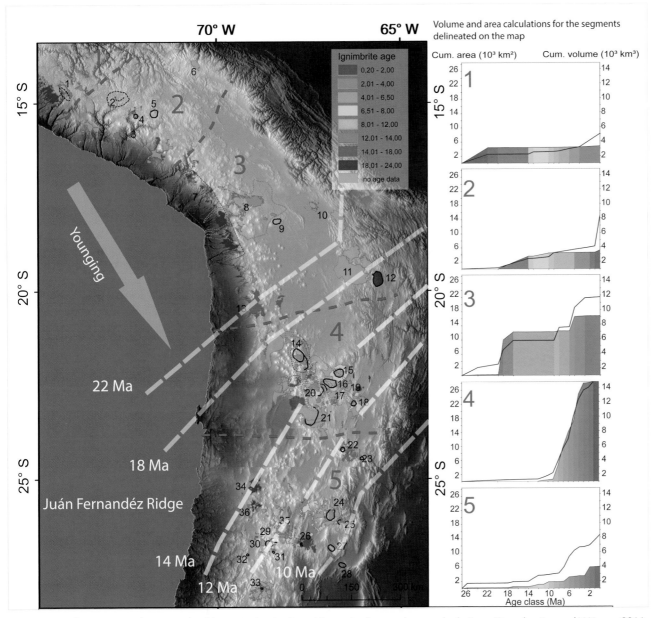

Overview of major ignimbrites and calderas in the Andes with results from volume calculations (Brandmeier and Wörner, 2016; Freymuth et al., 2015). Volumes of ignimbrites over time are calculated on the basis of our GIS aerial mapping and are combined with published data. Cumulative volumes (colored graph, "cumulative volume") and areal coverage (blue line graph, "Cumulative area") of ignimbrites are shown for each segment over the past 26 Ma in classes of 1 Ma. The shaded lines show the position of the Juán Fernández Ridge (Yáñez et al. 2001 and 2002). Modified from Freymuth et al. (2015).

The total minimum cumulative area covered by ignimbrites through time is 90,000 square km, and the corresponding minimum volume is 30,000 cubic km. Around 14,000 square km (500 cubic km) are from ignimbrites without a known age. This is a minimum estimate since it does not consider eroded areas but still shows the magnitude of these eruptions that we have never seen in historic times.

Magma production rates during ignimbrite flare-ups range between 21 and 86 cubic km per km arc per Ma. This estimate is in the same order of magnitude—but additional to—as the normal arc magmatic flux of 30 to 80 cubic km per km arc per Ma[-1] that is typical for the Andes. However, as these eruptions represent punctuated events in time, the impact and magnitude are enormous.

Combining these findings with compositional data analysis results in a better understanding of the overall evolution of the Andes and will be explained further.

THE BIG PICTURE

Timing of large-volume ignimbrite eruptions, their magma compositions, and eruptive volumes are closely linked to the thermal and structural evolution of the Andean crust. By combining and analyzing Andean-wide data, several striking patterns can be revealed and confirmed: volumetric calculations and spatial correlations show that ignimbrite flare-ups in the Central Andes follow the southward subduction of the buoyant Juán Fernández Ridge that is well known due to the publications by Yáñez and colleagues and explain a younging in eruption age. During ridge subduction, no volcanic activity takes place. This period is followed with a delay in massive eruptions (see the ridge subduction figure).

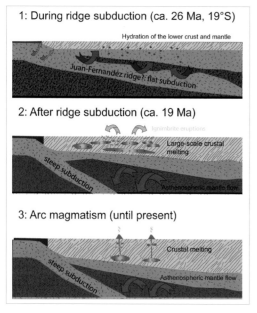

This temporal and spatial shift in super-eruptions produces systematic compositional changes related to crustal thickening and mountain building. The application of REE to geology is important to understanding the petrological processes of rock formations because they act as geochemical markers that can be used to infer the mechanisms that have affected a rock. Variations in major elements and REE patterns of large-volume ignimbrites with time and location from north to south show significant differences between the old, large-volume ignimbrites (e.g., Nazca, Oxaya, Huaylillas, ca. 22 to 13 Ma) in northernmost Chile and southern Peru and the young (ca. 10 to 2 Ma) Altiplano-Puna Volcanic Complex ignimbrites (e.g., Atana, Galán) with similar volumes. Cluster analysis results, when considered for different age and volume classes, indicate that old large-volume ignimbrites are more rhyolitic and less depleted in heavy rare earth elements (HREE) than young ignimbrites. Such systematic patterns in major elements and in REE, however, are not valid for small-volume ignimbrites because of their different origin in local small magma systems that do not involve large-scale crustal melting as proposed for ignimbrite flare-ups and, thus, super-eruptions.

When comparing mean values of centered log ratio transformation (Samarium, Ytterbium, Europium) for the respective age classes, we observe a geochemical transition zone between ca. 13–9 Ma (as shown in B in the graphic on eruption rates). This change in REE patterns is interpreted as a progressive shift in pressure-dependent mineralogy that indicates a thickening crust over time and therefore surface uplift.

The schematic illustration shown above shows the temporal succession of events as we envisage them exemplified for a section at 19° southern Latitude. In a first stage, while the ridge is subducted, no volcanic eruptions occur because no melting can take place. However, the crust is enriched in fluids from the ridge. In the second step, after the ridge has passed, hot asthenospheric mantle material gets into contact with the upper crust and triggers massive melting and eruptions. In a last stage, only normal arc-magmatism occurs until present times.

The advective heat and magma input into the crust (second step) occurred into differently "conditioned" upper Andean crust with respect to time and space: cold and thin crust in the north at 25 to 13 Ma and hot and thick crust in the south at <12 Ma. This process can explain different crustal signatures through time and space and might reflect the rise of the Andes. However, the overall crustal evolution in time and space is highly complex, and various processes at different crustal levels interact and are not completely understood, inviting further studies in this fascinating field of research. The GIS database we provide can be of great help in further studies exploring volcanism in the Central Andes. The possibility of having an integrated GIS system that allows mobile data collection as well as advanced statistical methods will hopefully lead to even more insights in volcanology and other fields of research in the future.

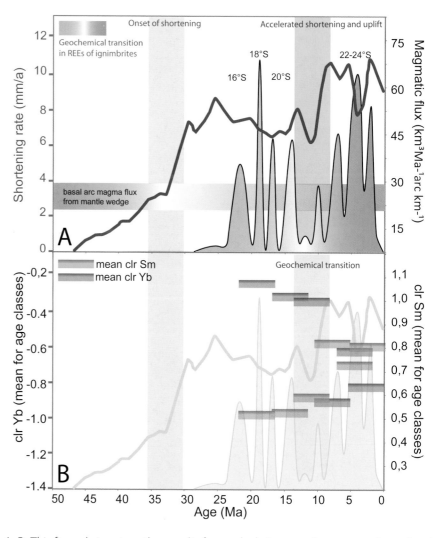

A–B: This figure brings together results from calculating eruption rates and geochemical analysis and combines them with crustal shortening rates that indicate the uplift of the Andes. In detail, the figure shows a comparison of magmatic flux and ignimbrite flux (assuming a volcanic:plutonic ratio of 1:5) and average shortening rates (Oncken et al. 2006; thick red graph in A). The second pulse of accelerated shortening coincides with a transition (ca. 13 to 9 Ma) between geochemical clusters toward lower clr Yb and higher clr Eu and clr Sm (B) parallel to crustal thickening. Note that pulses also migrate in space (north to south). Modified from Brandmeier & Wörner 2016.

ENDNOTES

Photos on pages 42–43 and 47 by G. Wörner.

Mount St. Helens photo courtesy of USGS.
Mount Tambora photo courtesy of the NASA Expedition 20 crew.
Eyjafjallajökull photo by Henrik Thorburn.

All other photos by Melanie Brandmeier.

References refer to the present text. Further references related to the database and mapped ignimbrites can be found in the database.

Brandmeier, M., and G. Wörner. 2016. "Compositional Variations of Ignimbrite Magmas in the Central Andes Over the Past 26Ma—A Multivariate Statistical Perspective." *Lithos* 262: 713–28.

Coira, B., et al. 1993. "Upper Cenozoic Magmatic Evolution of the Argentine Puna—A Model for Changing Subduction Geometry." *International Geology Review* 35, no. 8: 677–720.

De Silva, S., et al. 2006. "Large Ignimbrite Eruptions and Volcano-Tectonic Depressions in the Central Andes: A Thermomechanical Perspective." Geological Society, London, *Special Publications* 269, no. 1: 47–63.

De Silva, S. L. 1989. "Altiplano-Puna Volcanic Complex of the Central Andes." *Geology* 17: 1102–06.

Francis, P. W., et al. 1983. "The Cerro Galan Ignimbrite." *Nature* 301: 51–53.

Freymuth, H., et al. 2015. "The Origin and Crust/Mantle Mass Balance of Central Andean Ignimbrite Magmatism Constrained by Oxygen and Strontium Isotopes and Erupted Volumes." *Contributions to Mineralogy and Petrology* 169, **no. 6**: 58.

Kay, S. M., and B. L. Coira. 2009. "Shallowing and Steepening Subduction Zones, Continental Lithospheric Loss, Magmatism, and Crustal Flow under the Central Andean Altiplano-Puna Plateau." *Backbone of the Americas: Shallow Subduction, Plateau Uplift, and Ridge and Terrane Collision* 204: 229.

McDonough, W. F., and S. S. Sun. 1995. "The Composition of the Earth." *Chemical Geology* 120, nos. 3–4: 223–53.

Oncken, O., et al. 2006. "Deformation of the Central Andean Upper Plate System—Facts, Fiction, and Constraints for Plateau Models." *In the Andes: Active Subduction Orogeny*, edited by O. Oncken, G. Chong, G. Franz, et al., 3–27. Springer Berlin Heidelberg

Schmincke, H.-U. 2004. *Volcanism*. Berlin: Springer.

Wörner, G. U., D. I. Kohler, and H. Seyfried. 2002. "Evolution of the West Andean Escarpment at 18°S (N. Chile) During the Last 25 Ma: Uplift, Erosion and Collapse Through Time." *Tectonophysics* 345: 183–98.

Yáñez, G. A., et al. 2001. "Magnetic Anomaly Interpretation Across the Southern Central Andes (32–34 S): The Role of the Juán Fernández Ridge in the Late Tertiary Evolution of the Margin." *Journal of Geophysical Research: Solid Earth* (1978–2012) 106, no. B4: 6325–45.

Yáñez, G., et al. 2002. "The Challenger–Juán Fernández–Maipo Major Tectonic Transition of the Nazca–Andean Subduction System at 33–34 S: Geodynamic Evidence and Implications." *Journal of South American Earth Sciences* 15, no. 1: 23–38.

Hiking down from the Parinacota volcano.

PREDICTING GLOBAL SEAGRASS HABITATS

Using machine-learning techniques to study a mostly hidden but environmentally crucial marine resource, scientists are building geographically linked models that show where seagrasses are expected to flourish under differing ocean conditions.

By Orhun Aydin and Kevin A. Butler, Esri

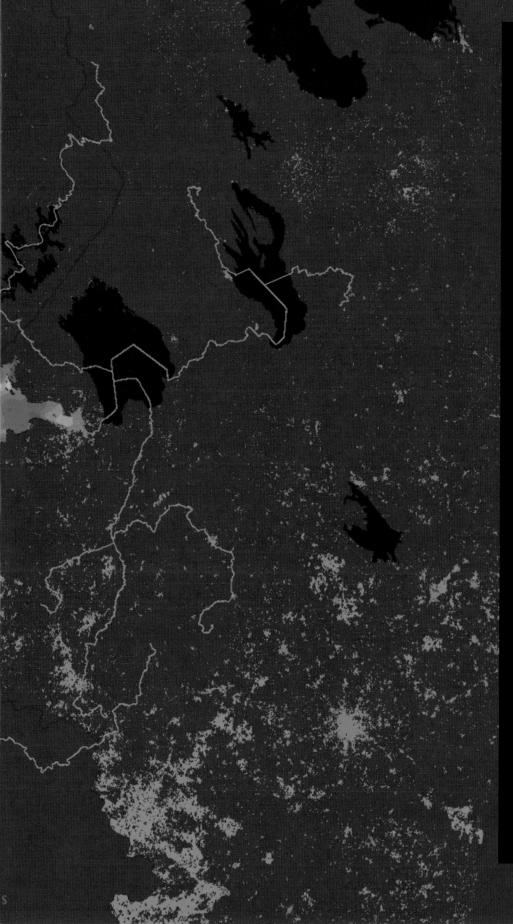

The art of telling stories with maps

Many of the sciences, especially those that deal with Earth systems and biology, have to do with *location*. Whether it's the stately grinding of tectonic plates or the interdependencies of plants and animals within ecosystems, location is key to discerning patterns and detecting trends. Just as essential to science is *communication*.

Beautiful multimedia stories

Increasingly, scientists are using the web to share their findings with broader audiences. The web enables instantaneous publication, frequent updating, and use of rich multimedia: text, images, video, infographics, and maps. But maintaining custom websites and managing multimedia content can be daunting. Enter ArcGIS® StoryMaps, a web app that demystifies multimedia publishing. Story maps enable the creation of beautiful multimedia stories without requiring specialized web development skills.

Government agencies, including the National Oceanic and Atmospheric Administration and the US Geological Survey, are making story maps to interpret scientific research for broad audiences. Academic departments in scores of colleges and universities are using story maps as instructional tools and are challenging students to create their own. NGOs are creating story maps as well: GRID-Arendal produced the story map on eutrophication in the Baltic Sea shown on these pages; it's one of a series of stories that enables GRID-Arendal to pursue its mission of "organizing and transforming available environmental data into credible, science-based information products."

Modes of persuasion

Aristotle's "modes of persuasion" provide a road map for storytelling that is useful for scientists today. *Logos*, the natural habitat of scientists, uses reason to persuade an audience. *Ethos*, conveying the authority and credibility of the persuader, requires scientists to defy their habits and place themselves in the narrative. *Pathos* might be the biggest stretch for scientists trained on objectivity. But all scientists have a passion for their work and can tap that passion in order to elicit an emotional response from their audiences.

Finally comes *Kairos*, bringing time and place into the story. It is place, of course, that makes story maps a uniquely useful storytelling platform. Maps add extra dimensions to multimedia stories. They pin a narrative to place. They place a story within a larger context. They provide additional, deeper insights.

Story maps are a way to help science become accessible and scientists to become storytellers.

Allen Carroll, Esri.

THE POWER OF STORYTELLING FOR SCIENCE

The map shows the gradient of nitrogen surface concentrations (0–10 m; 0–240 µmol L–1) in the Baltic Sea, with highest concentrations close to land and often in proximity to the mouth of major rivers, populated areas, and agricultural land. In the Baltic Sea region, 70–90 percent of the nitrogen originates from agriculture, and is then transported from the fields over the rivers and streams into the sea.

Agricultural land

Populated areas

Major rivers

Nitrogen (µmol/l)

240

0

Sources: CIESIN, HELCOM.

PARADISE LOST
Camp Fire 2018: Burn index using Sentinel-2 imagery

On November 8, 2018, at 6:33 a.m., a fire was reported near Poe Dam and Camp Creek in northern Butte County, California. Driven by 50 mph winds and dry conditions, the rapidly growing wildfire rolled seven miles west into the rural town of Paradise at 8 a.m., destroying most of the community. Eventually, the Camp Fire consumed more than 150,000 acres, burned more than 18,000 buildings, and killed more than 85 people in the deadliest wildfire in the United States since 1918.

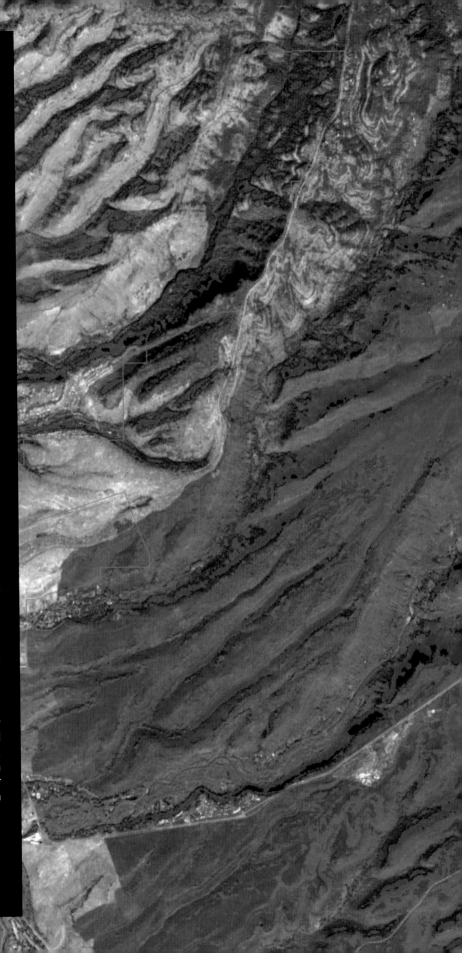

SENTINEL-2 IMAGERY VIEWER

Dramatic events like the Camp Fire that nearly destroyed the California town of Paradise are recorded by satellite platforms that continuously orbit Earth. The satellites deliver a stream of imagery and remote-sensing data that helps people understand how the earth changes over time. One of these systems, Sentinel-2, is a set of twin satellites flying in the same orbit 180 degrees apart. The mission of Sentinel-2 is to collect wide-swath, high-resolution, multispectral imagery. The image data is used to manage land, agriculture, and forestry but also contributes to emergency management, humanitarian relief operations, and global security.

Sentinel-2 is part of Copernicus, the world's largest single Earth-observation program directed by the European Commission in partnership with the European Space Agency (ESA). Sentinel-2 provides multispectral data spanning 13 bands in the visible, near-infrared (NIR), and short-wave infrared with spatial resolution ranging from 10 meters to 60 meters. Sentinel-2 continuously collects and provides imagery for any land-based location in the world every five to seven days.

Imagery exploration and analysis

Sentinel-2, Landsat, and a vast library of contemporary and historical imagery is available through the ArcGIS® Living Atlas of the World. ArcGIS software and services provide additional access to enriched visualization, geospatial analytics, and image exploitation tools.

This image was produced from ESA Remote Sensing data with the Sentinel Explorer web app. Sentinel Explorer—accessible at sentinel2explorer.esri .com—provides quick access to many imagery band combinations, such as natural color, NIR, SWIR, and other indices for visualizing vegetation, moisture, and burn scars on the landscape. Sentinel Explorer allows users to perform change detection between two images of different dates, visually via a slider tool, or through image analysis.

Wildfire burn area assessment

Using Sentinel Explorer to compare imagery taken a few days before the fire to imagery taken shortly after the fire's containment, it's possible to generate a burn index (shown in orange). The burn index displays the detailed burn pattern of the Camp Fire, which spread over mountain ridges, through valleys and arroyos, and down the western slopes of the Sierra Nevada range into valley communities. Default rendering of the normalized burn ratio (NBR) is computed using multiple bands including NIR band 8 and SWIR band 12. With Sentinel Explorer, this display of band combinations and calculated indices is done on the fly within the web app. In this case, the burn index overlays the natural color imagery taken by Sentinel 2 several days after the fire was contained. The town boundary of Paradise is shown for reference.

Sentinel Explorer: Peter Becker, Esri
Data source: Copernicus Sentinel data for November 6, 2018, and December 6, 2018. This map contains modified Copernicus Sentinel data for November 6, 2018, and December 6, 2018.

A HIGH-RESOLUTION MARTIAN DATABASE

HiRISE imaging for exploring Mars

The High-Resolution Image Science Experiment (HiRISE) camera onboard the Mars Reconnaissance Orbiter (MRO) platform has collected remarkable images of the Red Planet since November 2006. Mission planners at the National Aeronautics and Space Administration (NASA) and Jet Propulsion Laboratory (JPL) use these images to select landing sites for rover-based and stationary missions, such as the InSight Mars Lander that touched down November 26, 2018. HiRISE was built by Ball Aerospace & Technologies Corp. under the direction of the Lunar and Planetary Laboratory (uahirise.org) at the University of Arizona. All available HiRISE images have been released at full resolution hosted on the cloud as a tiled service. Links to these publicly accessible services can be found at GISforScience.com.

Strange beauty

HiRISE images offer intriguing and detailed research material for numerous locations on Mars. In addition to mission planning, researchers can use the long, narrow strips of imagery to study surface features as small as one meter in size. HiRISE reveals the stark and strange beauty of sand dunes, impact craters, ravines, and canyons. These images help scientists and researchers explore Mars, search for indications of water, and even track how the planet is changing.

Sense of scale

Here, the HiRISE observation strip is located within the Aram Chaos crater, a large, ancient, and highly eroded impact crater. At more than 280 kilometers (174 miles) across, Aram Chaos would engulf the entire Los Angeles basin and San Diego. Zooming in reveals many smaller craters, including the one shown in the larger image. Like waves in a tiny ocean, sand dunes line up across the base of the crater. Clearly visible are apparent rock debris and possible landslides along the inside and outside rim.

Los Angeles

San Diego

HiRISE image processing and rendering: Lucian Plesea, Esri

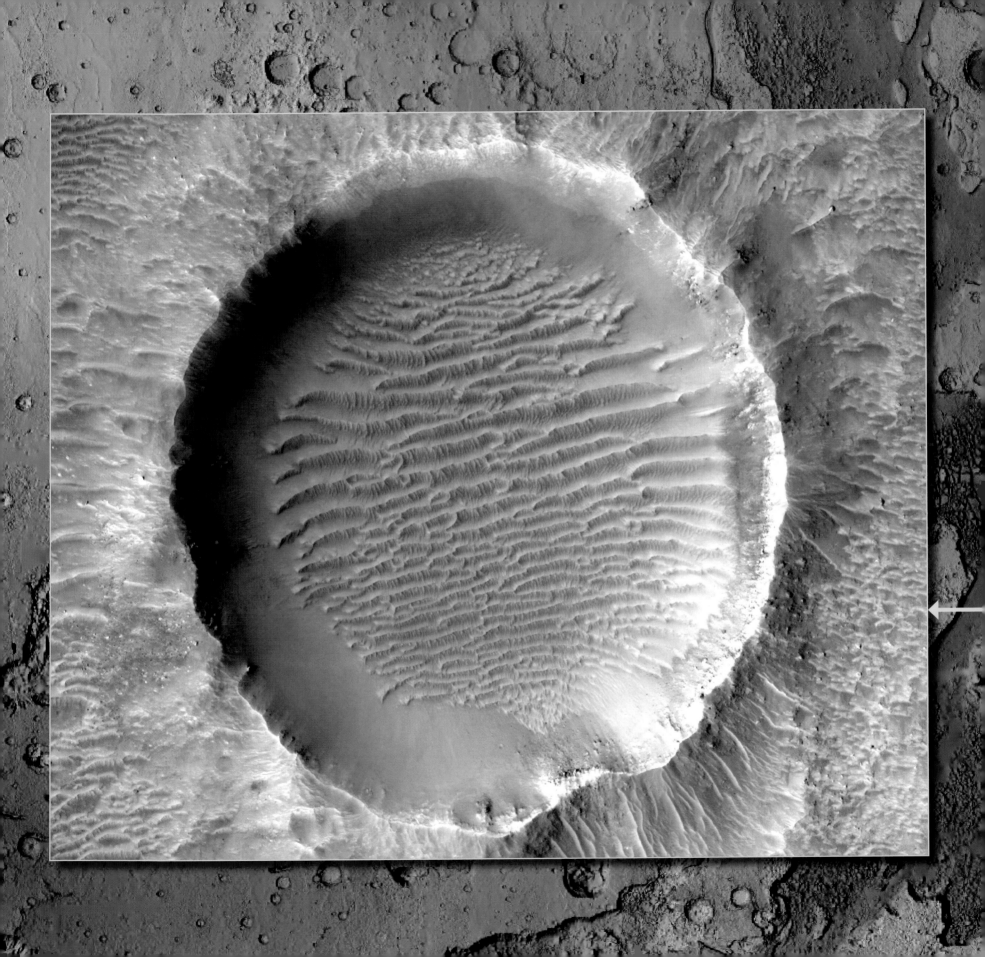

In this 3D model, the map shows the highest water depth in a simulated 24-hour period for Onion Creek near Austin, Texas, during a flood. Onion Creek is Austin's largest watershed and is particularly vulnerable to flooding. The 3D map uses the National Water Model stream flow forecast.

NATIONAL WATER MODEL

A geoscience resource that informs emergency response

In the United States, floodwaters kill more than 80 people a year and cause nearly $8 billion in losses. During the past five years, the National Oceanic and Atmospheric Administration (NOAA), in collaboration with University Corporation for Atmospheric Research (UCAR), Esri, and the academic research community, developed a new stream flow forecasting system, known as the National Water Model. The model, driven by approximately 7,000 observational measurements, hourly precipitation forecasts, and landscape characteristics, estimates water flow on 2.7 million stream reaches across the continental United States. The national model is a collection of models forecasting from hourly to 30 days in the future.

Flooding forecast maps

Emergency management planners and analysts use the hydrologic model to transform stream flow into water depth. An ArcGIS workflow using a digital elevation model and ancillary information turns water depth into flood forecast maps. These maps help users more accurately predict how a given storm might flood different areas, determine when the peak flood will occur, and estimate how long it will take to recede. This information is crucial for first responders and emergency management commanders. Having this predictive capability from hours to days before a flood improves their ability to plan for evacuations, road closures, and rescues.

Exploration in 3D

With this ability to predict when and where flood waters will strike, users can overlay this information with map layers showing buildings, roads, and demographics. Having all this information brought together in a map helps first responders understand the evolution of a flood emergency. In this example, an interactive 3D web application allows planners and responders to explore where water may accumulate during the next 24 hours. With this information, responders can set evacuation areas, route evacuees, position equipment, and visualize the impact of floodwaters on specific buildings and roads.

Modeling the event

The flood map is modeled on a 1-meter resolution, lidar-derived elevation model. The model computes the affected population using the Esri GeoEnrichment Service on the ArcGIS platform for the flooded area. Estimated damage costs are computed for each structure based on square footage and maximum depth of flooding derived from a Federal Emergency Management Agency (FEMA) table.

Estimated Cost: $28,292,020

People, property, and infrastructure affected:
3,169 people
1,242 total structures
987 residential structures
255 commercial structures
85 roads

3D flooding forecast: Steve Kopp, Esri
National Water Model contributors: National Oceanic and Atmospheric Administration,
University Corporation for Atmospheric Research, Esri, and numerous academic and
federal research partners. Visit water.noaa.gov.

Potential Loss
$50,000
$25,000
$5,000

Predicted Dissolved Oxygen Levels

Highest — 6.35 micromoles/liter

3.35 micromoles/liter

Lowest — 0.35 micromoles/liter

Scientists and researchers often encounter the problem of unknown values within a range of known values. To solve this problem, they use mathematical and statistical methods to create or interpolate new data points. They use these data points to fill in the gaps and model phenomena occurring across a landscape or in 3D space. Researchers use interpolation to accurately predict values for new data points using the existing values of a limited number of sample data points where the measurements were gathered. Researchers have begun using Empirical Bayesian Kriging 3D, a geostatistical interpolation method developed by Esri, to get new estimated values between the known values, and then model the measurements in 3D.

Kriging

Kriging is a method of interpolation used in spatial analysis in which data points exist at specific locations in space and time. Each data point is defined by geographic coordinates (i.e., latitude, longitude, and often, elevation) and other measurements, such as the amount of particulate matter in the air. Kriging applies the basic principle that distributed data points are spatially correlated. This principle assumes that while everything is related to everything else, near things are more related than distant things.

Researchers can use kriging to predict unknown values for any data point, such as elevation, rainfall, chemical concentrations, and noise levels. Empirical Bayesian Kriging 3D is used when the data points are distributed within a geographic volume, such as a square-mile study area of an ocean, ranging from the ocean surface to the ocean floor.

Cross validation

Empirical Bayesian Kriging 3D provides cross-validation tools to assess how well the model predicts values at unknown locations. Cross validation removes a measured point and then uses all remaining points to go back to the location of the removed point. This process is called the "leave-one-out" validation method. The measured value at the hidden point is then compared to the prediction value from cross validation. The difference between these two values is called the *cross-validation error* and is performed on every input point.

Modeling dissolved oxygen levels in Monterey Bay

Recent research has established that global oxygen levels in the ocean have declined for decades. Dissolved oxygen is the essential ingredient for life beneath the surface of lakes, rivers, and oceans. Using data from the World Ocean Database (WOD) provided by NOAA's National Centers for Environmental Information, researchers measured the levels of dissolved oxygen by sampling water at different locations in Monterey Bay on the California coast. From above, the sample locations would appear as dots on a map riding on a flat surface. However, researchers sampled at multiple depths at each location, leaving lots of distance between measurements, where dissolved oxygen levels are unknown. Researchers filled in the blanks using Empirical Bayesian Kriging 3D to interpolate the values between these known measurements and create estimated but reliable measurements at each sampling depth.

The result is a 3D map layer stack of surface models, each depicting ranges of dissolved oxygen levels. It is easy to see that dissolved oxygen levels vary across each surface and vertically through the layers. Cross-validation tools and charting allow researchers to explore any location across each surface and slide up and down through the surface layer stack at a specific location.

3D EMPIRICAL BAYESIAN KRIGING

Modeling dissolved oxygen in Monterey Bay

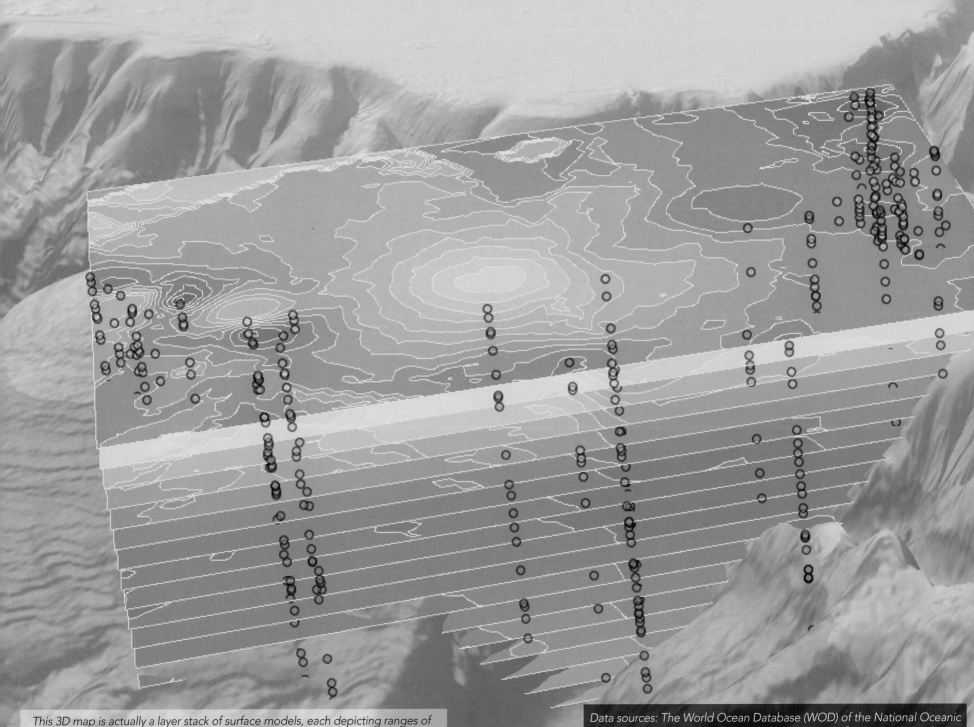

This 3D map is actually a layer stack of surface models, each depicting ranges of dissolved oxygen levels at various depths in Monterey Bay off the coast of Northern California. It is easy to see that dissolved oxygen levels vary not only across each surface but also vertically through the layers.

Data sources: The World Ocean Database (WOD) of the National Oceanic and Atmospheric Administration National Centers for Environmental Information (NOAA NCEI); Esri World Ocean Basemap; Esri World Elevation 3D/TopoBathy 3D. Modeling: Eric Krause, Esri.

LANDFALL:

A pythonic analysis of more than 160 years of hurricanes and cyclones and their impact around the Gulf of Mexico and Caribbean

SAFFIR-SIMPSON HURRICANE WIND SCALE
IN THIS MAP, THE HIGHEST RANKING OF A HURRICANE APPROACHING LANDFALL

- CATEGORY 5 (WINDS: 157 MPH OR HIGHER)
- CATEGORY 4 (WINDS: 130–156 MPH)
- CATEGORIES 1–3 (WINDS: 74–129 MPH)

Modeling: Atma Mani, Esri
Data Sources: National Hurricane Center, NOAA

Gulf of Mexico

REPUBLIC

PUE

JUPYTER™ NOTEBOOK ANALYSIS

The National Hurricane Center has collected meteorological data about hurricanes and cyclones across this planet from the past 160+ years. Esri recently analyzed this massive collection of datasets based on hurricane severity indicators, such as wind speed, atmospheric pressure, track duration, track length inland, and category over time (seasons). The analysis revealed that the total number of hurricanes is decreasing globally since 1970 across all basins. While the number of hurricanes shows a reducing trend, their severity is increasing, with hurricane wind speed, track duration, and category designation correlating positively and with atmospheric pressure correlating negatively against seasons. The research found a reduction in the number of milder, Category 1 and 2 hurricanes and an increase in the number of severe, Category 4 and 5 hurricanes. This map shows a detail of the worldwide hurricane landfall tracks analysis within the geographic region of the Gulf of Mexico and the Caribbean.

By analyzing the datasets as a time series, Esri observed a sinusoidal seasonality trend indicating that the peaks between hurricane events in the Northern Hemisphere and cyclones in the Southern Hemisphere were offset by about six months, matching the time when summer occurs in these hemispheres. One difference is that hurricanes over the Indian basin occur throughout the year because they are influenced by a monsoon phenomenon.

When Esri overlaid the hurricane tracks over coastlines, the trend showed that most hurricanes make landfall. Once they do, most of them travel less than 100 miles inland. Using spatial density analysis on the landfall locations, Esri found that repeat landfall of hurricanes affects the Atlantic coast along North and South Carolina in the United States, the states of Odisha and West Bengal in India, several areas on China's east coast, the southern tip of Japan, and most of the Philippines. By using GeoEnrichment, the process of adding demographic and lifestyle data to maps, we get a sense of how hurricane landfalls may impact people and places, although more research is needed.

Massive datasets such as the one used in this study don't typically fit within the computational memory of a scientist's laptop or workstation and are quite common in climate studies. To solve a problem of this scale, Esri used distributed and delayed analysis tools with ArcGIS API for Python to wrangle and aggregate the data. The analysis was performed entirely in Python on ArcGIS Notebooks, a hosted Jupyter Notebook environment that provides access to datasets and tools from Esri and many analytical libraries available in the scientific Python

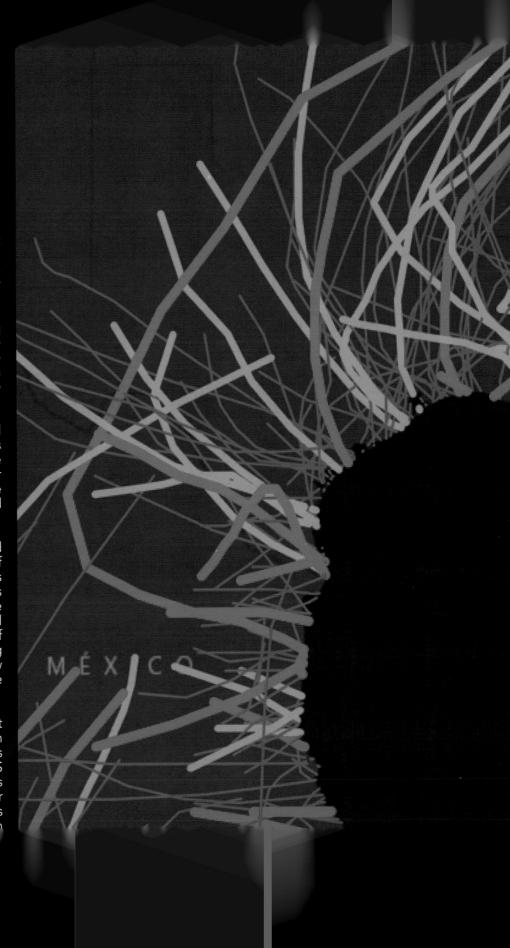

MÉXICO

As the world changes, connected landscapes are more likely to survive than isolated environments. A nationwide green infrastructure model enables conservation groups, nonprofits, businesses, and government agencies to focus on connected open spaces. Their goal is to preserve farmland, habitat, and water quality and promote outdoor recreation, which improves community health and supports local economies.

Green infrastructure resources available from the ArcGIS Living Atlas of the World provide a collection of data and tools to help agencies analyze habitats identified as unfragmented natural areas (also known as *cores*).

Cores

Cores, larger than 100 acres in size and at least 200 meters wide, are largely undisturbed areas. Maps of these areas include hydrology, species, form, elevation, soils, and ecosystem-related data. The process of creating the national model identified more than 570,000 core areas across the contiguous United States. The cores model generates a high-quality map depicting most intact habitats.

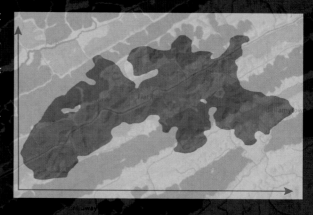

In this map, a set of cores that intersect within the immediate vicinity of Middlesboro, a city in Bell County, Kentucky, are selected. Middlesboro is one of the few cities in the world situated inside what geologists believe is an ancient meteorite crater, or astrobleme. In Middlesboro, this geological phenomenon helps accentuate the arrangement of green infrastructure cores surrounding a human-modified landscape.

Planning

Green infrastructure planning involves the identification and protection of intact habitat cores. The process includes identifying an area's natural assets for protection, and then determining which natural assets are most critical to the local area. The process starts by identifying large, intact landscape cores. Each core contains more than 50 unique attributes related to its characteristics, of which 21 can be used with the filtering app to create a customizable selection of cores based on variables of interest. By using a built-in weighting model, researchers can generate a ranking score for each core.

Once cores are selected, researchers and planners can access and interrogate the science behind the green infrastructure model by filtering variables associated with the selected set. For example, one such variable is called *human modified mean*, which measures the degree of human impact upon the landscape. The index ranges from 0.0 for a virgin landscape condition to 1.0 for the most modified areas. The average value for the United States is 0.375.

Scoring

Local scoring allows for the ranking of cores for different projects to make the best use of the national data. Scoring the cores allows planners to further filter out the cores that may not be as important to their project. Applying model weights to each characteristic generates new scores ranking each core. Model weights can be adjusted to emphasize cores with conditions conducive to a project.

Modeling: Ryan Perkl, Esri.

Pipefish and tube-snouts (Aulorhynchus flavidus) in an exhibit with seagrass and seaweeds at the Monterey Bay Aquarium.

ECOLOGICAL VALUE OF SEAGRASS

Seagrasses are a group of bottom-attached, flowering plants that have adapted to exist fully submersed in salty and brackish coastal waters.[1] They have roots, stems, and leaves and produce flowers and seeds. While often confused with seaweeds, they are more closely related to terrestrial flowering plants and often grow in large patches or meadows known as *beds*. They are found along the coasts of every continent except Antarctica.

Meadows of seagrass serve as the foundation of highly productive ecosystems and provide food and shelter for a variety of marine life, ranging from invertebrates to marine mammals. They help stabilize the ocean bottom and maintain water quality. Most importantly, seagrasses play an important role in combating global warming, storing up to 100 times more carbon dioxide per unit area compared with tropical forests. Seagrass habitats are declining globally because of climate change and human impacts such as eutrophication (excessive nutrients in the water from agricultural fertilizer runoff); sediment discharge from the land, which reduces the amount of sunlight available; and dredging or boating activities that physically uproot the plants.[2] Despite the benefits and importance of seagrass, little is known about the global distribution and diversity of this valuable resource.

Types of seagrass

At least 50 species of seagrass from four families (Posidoniaceae, Zosteraceae, Hydrocharitaceae, and Cymodoceaceae) exist.[3] They have a variety of leaf shapes ranging from thin tubes to wide, oblong paddles and range in size from a few inches to 35 feet. Although the benefits of seagrasses as a whole are generally well accepted, the degree to which specific benefits and services are provided by individual species has not been quantified.

Seagrass species vary in size and leaf shape. Above, paddle grass (Halophila decipiens) is found in tropical regions of the Indian and Pacific Oceans, the Western Atlantic Ocean, and European waters.

Turtle grass (Thalassia testudinum) found in calm, shallow waters throughout the Caribbean Sea and the Gulf of Mexico and as far north as Cape Canaveral in Florida.

Squids, octopuses, and cuttlefish, such as this Australian giant cuttlefish (Sepia apama), make up just one group of animals that call seagrass beds home.

STATUS OF GLOBAL SEAGRASS ECOSYSTEMS

Seagrasses provide a variety of ecosystem services, functions that benefit humans either directly or indirectly. Putting a monetary value on ecosystem benefits is difficult and controversial, but marine scientists generally agree that seagrasses form the foundation of marine coastal ecosystems.[4]

Seagrass meadows provide habitat to fish and infauna, creatures that live in the soft sediments of the seafloor, serving as refuge from predators. Seagrasses mitigate the impacts of strong currents, and in the case of creatures living in the sediment, seagrass roots make it more difficult for predators to dig.

Seagrass plays a major role in maintaining water clarity, an important part of marine ecosystem health. Clear water allows sufficient sunlight to penetrate the water column and support productivity. Clear water allows prey to more easily spot and avoid predators and can increase the efficiency of foraging. Seagrasses capture and hold fine sediments and reduce the sediment stirring caused by waves and currents that would otherwise create turbid waters.

Juvenile fish and invertebrates thrive in the "nursery" habitat of seagrass meadows partly because of lower predation risk. A common presumption has been that nursery habitats result in faster growth of juveniles because of the higher abundance of food in seagrass meadows. However, this assumption has been questioned.[5]

The leaves and roots trap a relatively small amount of total carbon. Comparativly, sediments under the seagrass meadows capture most of the carbon.[6] As seagrass and the animals they host die, their carbon is trapped in sediments that raise the seafloor by as much as one millimeter per year. These sediments can accumulate for thousands of years, trapping large amounts of carbon.

Threats to seagrass habitat

A comprehensive review of 215 studies revealed that seagrass meadows are disappearing at a rate of approximately 110 square kilometers per year globally. Overall, 29 percent of all seagrass area has been lost since 1879, when measurements of seagrass extent were first recorded.[7] Because seagrasses grow primarily in shallow coastal areas, they are vulnerable to impacts from human activities in often heavily populated areas. For example, nutrient-rich runoff from agricultural activities can produce eutrophication, resulting in less available light for plant growth. Increased urban development along coasts results in increased sediment runoff, creating turbid waters that reduce the amount of available light. Seagrasses can also be disturbed by physical uprooting from boating and other recreational activities.

Arguably the most globally extensive threat to seagrasses is climate change. Seawater temperatures impact seagrasses. Reynolds et al. (2016) summarize the impact of increased water temperatures on seagrass, stating, "Rising water temperatures tend to increase rates of seagrass respiration (using up oxygen) faster than rates of photosynthesis (producing oxygen), which makes them more susceptible to grazing by herbivores. Increased temperature also increases seagrass light requirements, influences how quickly seagrasses can take up nutrients in their environment, and can make seagrasses more susceptible to disease."

Seagrasses provide important nutrition for many marine species, including this Green Sea Turtle grazing at Akumal Bay in Yucatan, Mexico.

UNDERSTANDING WHERE SEAGRASS WILL GROW

As with all living organisms, seagrasses need specific environmental conditions to thrive. Unlike animals that can migrate to new locations when climate conditions become unacceptable, bottom attached seagrasses have few adaptations to climate change. Understanding how seagrass will respond to a changing climate first requires an understanding of the physical and chemical properties of current seagrass habitats. Forecasting future seagrass habitats requires these steps:

1. Modeling the relationships between seagrass habitats and current coastal water conditions
2. Establishing a relationship (baseline) between ocean temperature and other variables such as salinity and nutrients
3. Using the baseline model, simulating future ocean conditions given incremental increases in ocean temperature
4. Modeling seagrass habitats for future ocean conditions

First, we established the relationships between seagrass occurrence and ocean conditions. We approach seagrass habitat modeling as a binary classification problem where a 1 indicates the presence of seagrass and zero indicates the absence. Using a random forest model, we determined the ocean conditions (e.g., temperature range, salinity level, etc.) necessary for seagrass presence. The model also determines the ocean conditions where seagrass is absent. To simulate realistic future ocean conditions, we first established the present-day relationships between temperature and other ocean variables such as salinity and nutrients. This step allowed us to simulate ocean conditions under increasing temperatures. For example, how will nutrient levels change if ocean temperatures increase by 0.2°C? We simulated 20 "future" oceans by incrementally increasing ocean temperature by 0.1°C. The final simulation represents a 2°C increase as anticipated in the Kyoto protocol.[8] Finally, we predicted future seagrass habitats using the initial binary random forest model and the "future" ocean conditions.

We perform the workflow in ArcGIS® Pro with native machine-learning functionality and also external Python® libraries, integrated as Python toolboxes. We used this ArcGIS Pro model to prepare data sources for analysis and create a model for seagrass occurrence.

Workflow and data sources

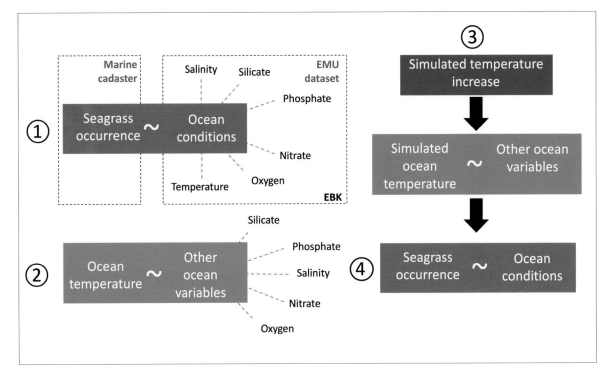

The chart depicts data sources and machine-learning methods used for modeling seagrass habitats.

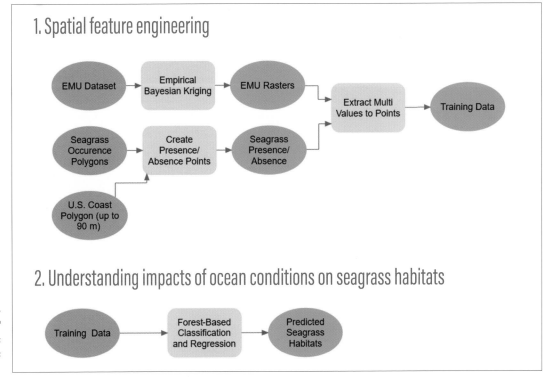

DEPLOYING MACHINE LEARNING

We use an unsupervised machine-learning method to create spatially contiguous regions based on the six physiochemical ecological marine unit variables. In the context of EMUs, a region is a spatially contiguous zone where the values of the EMU variables within the zone show little variability. Each region is as similar as possible based on all six EMU variables, and all the regions themselves are as different as possible. Generic machine-learning methods that perform similar tasks on nonspatial data fall under the umbrella of clustering methods. Clustering methods often need the number of regions to be specified up front.

This requirement is also true for the spatial counterpart of clustering, regionalization.[9] We use an explicitly spatial machine-learning method, SKATER,[10] to define neritic regions based on the EMU data. We employed a GIS-based implementation of the SKATER method that automatically detects the optimal number of regions using the pseudo F-statistic.[11] The SKATER method found six distinct neritic regions, which we labeled as landlocked seas, southern coasts of the Americas, coastal Antarctica, northern coasts, warmer and nutrient-rich coasts, and the Asia portion of the Ring of Fire, which includes hundreds of mostly underwater volcanoes situated in a horseshoe shape in the Pacific Ocean. Within each of these regions, we explored the relationships between temperature and the other five variables. The model in step 3 helps us delineate ocean regions and explore these relationships.

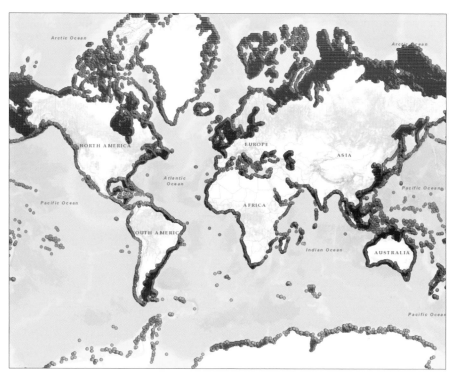

This world map identifies the six neritic regions of the world.

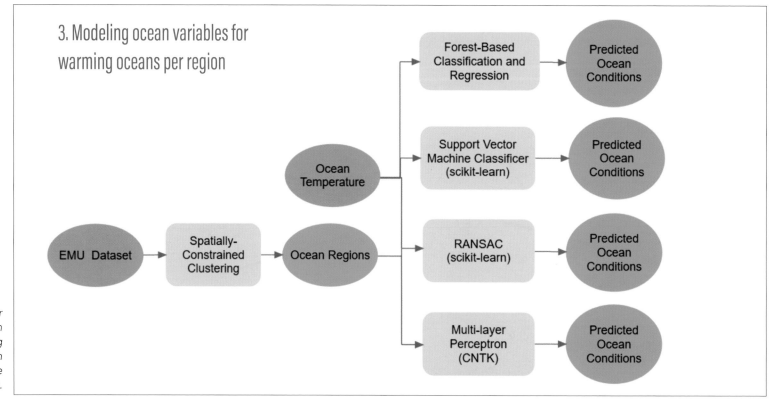

3. Modeling ocean variables for warming oceans per region

EMU Dataset → Spatially-Constrained Clustering → Ocean Regions

Ocean Temperature

- Forest-Based Classification and Regression → Predicted Ocean Conditions
- Support Vector Machine Classificer (scikit-learn) → Predicted Ocean Conditions
- RANSAC (scikit-learn) → Predicted Ocean Conditions
- Multi-layer Perceptron (CNTK) → Predicted Ocean Conditions

ArcGIS Pro model for defining data-driven ocean regions and exploring different regression models to predict future ocean conditions.

USING MACHINE LEARNING TO MODEL COMPLEX RELATIONSHIPS

For each of the regions, we built regression models to explore the relationship between water temperature and the other EMU variables. The scatterplot matrix showed that these relationships are complex, and it was not clear which machine-learning technique would best capture them. We investigated the following regression models using open-source Python machine-learning packages:

1. Random sample consensus (RANSAC)
2. Support vector machine (SVM)
3. MLP (Multi-layer perceptron)
4. Random forest (RF)

RANSAC forms a linear relationship between variables by fitting lines to random subsamples of the data. This method defines a final linear regression that is a consensus but is prone to outliers in data. SVM can capture complex relationships between temperature and other EMU variables in the neritic zone. Similarly, MLP can capture complex relationships; however, the algorithm typically captures overarching relationships, avoiding overfitting to data. Finally, random forest can model nonlinear relationships. Scatterplot graphics for one of the relationships, temperature versus salinity, show how well each machine-learning regression method captures the regional relationships.

Note that regions based on EMU variables allow us to divide the data spatially and to create contiguous, parsimonious models. We repeat the analysis above for each of the EMU variables and define regional regression models with temperature as the independent variable.

It is important to assess the quality of any analysis method, and machine-learning methods are no exception to this rule. For each of the regression methods, we analyzed the percentage of variance explained by the model for each region. Overall, RF explained a higher amount of variance compared with SVM, RANSAC, and MLP with a lower estimation error.

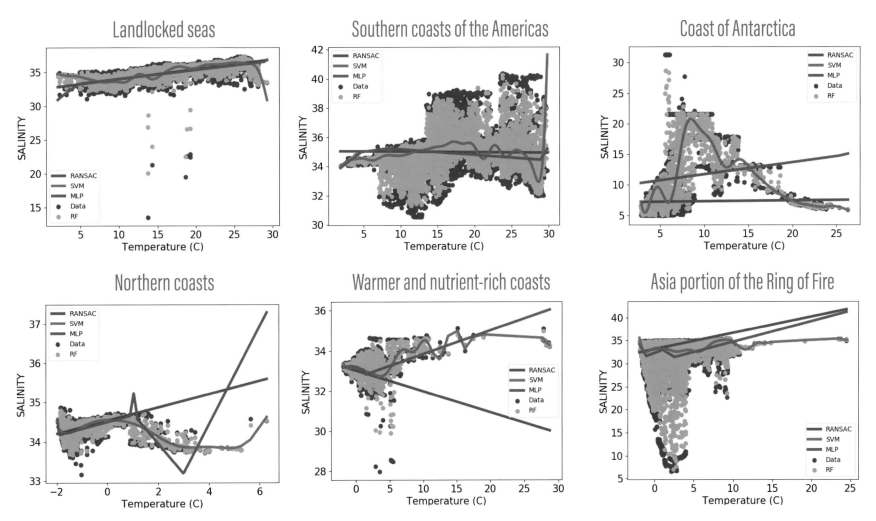

This six-part figure shows that SVM overfits to data, whereas RANSAC performs well in regions where the relationship is linear; however, it is not representative for nonlinear relationships. MLP is prone to oversimplify to data as we use a shallow network. Lastly, forest-based classification and regression (RF) can model complex relationships between temperature and other ocean variables. Thus, RF is used to model ocean variables for warming oceans.

MODELING PRESENT SEAGRASS HABITATS

To predict seagrass habitats under changing ocean conditions, we first built a baseline random forest model that predicts seagrass occurrence under current conditions. Our model is a binary classification where a 1 indicates the presence of seagrass and a 0 the absence. Here we are using random forest as a classifier, where the goal is to place each habitat into one of two categories, present or absent. We chose to model the presence/absence of seagrass habitats using a random forest classifier because of its ability to model complex combinations of different variables and predict a binary result. Random forest is a supervised machine-learning method. It needs training data so that it can "learn" what conditions promote or discourage seagrass habitat. We trained the RF model using seagrass occurrence data (the ones in our model) for the coastal United States. We downloaded the seagrass occurrence data from US Marine Cadastre. We evaluated the performance of the classifier by comparing its results (presence/absence) to a random guess of whether seagrass exists in an area. The receiver operating characteristic (ROC) curve graphic summarizes the model's performance.

An ROC curve is a metric of the ratio of locations that are correctly classified as having seagrass present versus areas that are incorrectly classified as having seagrass present, or false positives. The dashed line shows the success rate for random assignment or guessing. The ROC shows that our random forest classifier built for US coastal areas has high predictive power. We apply the model trained for the US coast to the global dataset. We aggregated the results, the number of times the model predicted seagrass presence, to hexagons and converted the counts to densities for easier visualization. The resulting map shows high densities of seagrass occurrence in Australia, Polynesia, and the North Sea. This map will serve as the baseline for comparison of seagrass habitat loss or gain under changing ocean temperatures. Our results are consistent with known, well-established seagrass habitats as reported by Short et al. We note that our model made predictions based on training data from the US coast only.

After numerically and empirically validating the seagrass habitat model, future seagrass habitats are predicted under increasing ocean temperatures. Count is a proxy for habitat suitability of an area within a unit volume (hexagon cell). If within a unit area more locations are predicted to be suitable for seagrass habitats, this area is a more suitable location for seagrass growth.

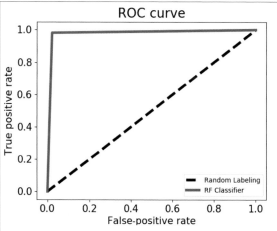

ROC curve for random forest classifier model for seagrass abundance.

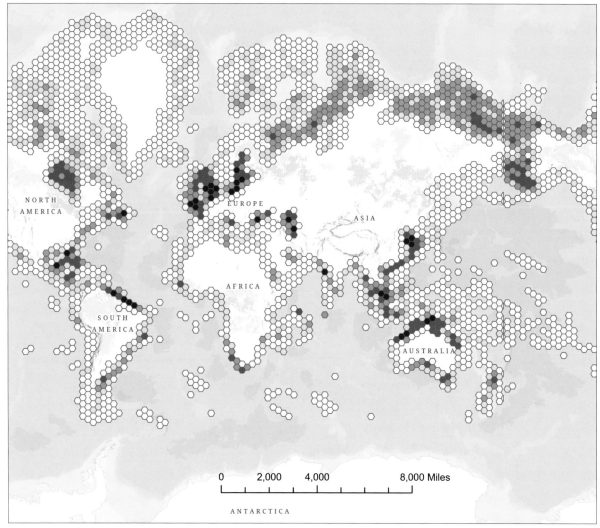

This density map shows the predicted seagrass abundance based on a random forest model trained for the US coast.

THE RELATIONSHIP BETWEEN SEAGRASS OCCURRENCE AND OCEAN CONDITIONS

Modeling future seagrass habitats involves simulating an increase in global ocean temperatures and estimating the resulting changes on other ocean characteristics such as salinity and nutrient content. We used these future ocean conditions in a presence/absence habitat model to predict future seagrass habitats.

Seagrasses live in the shallow portions of the world's oceans because they depend on sunlight. Coastal waters, also known as the *neritic zone*, respond rapidly to changes in atmospheric temperature. Thus, changes in atmospheric and oceanic conditions impact seagrass habitats. Seagrasses quickly respond to changes in their surroundings and function as "coastal canaries."[12] The decline in seagrass habitats is a warning sign that the ocean's health is in trouble.

Global climate change will alter the chemistry of the neritic zone by 2030.[13] Increasing temperatures in shallow waters change the solubility of nutrients and oxygen needed for seagrass growth. The graphic of average global sea-surface temperature from 1880 to 2015 shows a strong upward trend. While this global trend is informative, the oceans are dynamic, and we argue that the impacts of increasing temperature might manifest themselves differently in different regions. Later in this section, we implement regional models to understand the relationships between temperature and other ocean conditions.

We simulated 20 future oceans (neritic zone) by increasing current ocean temperatures by 0.1°C in each iteration, resulting in the final simulation being 2°C warmer than current conditions. To represent present-day ocean conditions, we extracted data for the neritic zone from the globally extensive ecological marine units (EMU) dataset.[14] The following data used in the habitat model represent physical and chemical characteristics of the ocean that drive the distribution of seagrass habitats (where °C is degrees Celsius and μmol is the symbol for the micromole):

1. Temperature (°C)
2. Salinity (μmol/l)
3. Dissolved oxygen (μmol/l)
4. Phosphate (μmol/l)
5. Nitrate (μmol/l)
6. Silicate (μmol/l)

The solubility of nutrients (phosphate, nitrate, and silicate) strongly depends on temperature. For each of the 20 future ocean temperature scenarios, using the simulated temperature, we need to calculate the values of the remaining five variables. We need to perform these calculations regionally because a visual exploration showed meaningful regional differences in the relationships between temperature and the other variables. For example, consider the global scatterplot of salinity and temperature. Overall, there is no clear relationship. However, interactive exploration using a dynamically linked scatterplot and map of the EMU data revealed regional trends. The scatterplot matrix graphic shows some of these unique regional relationships and strengthens the argument that a global model is not sufficient to capture the overall complexity.

The challenge we now face is, which regions should we use and what is the best method to capture the relationships between temperature and other ocean conditions?

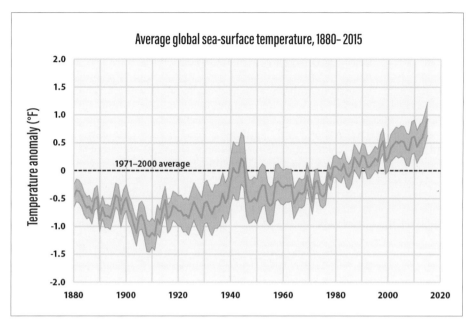

Average global sea-surface temperature in the past 135 years. Note the impact of human activity, such as the Second World War, and the increasing trend since the early 1900s.

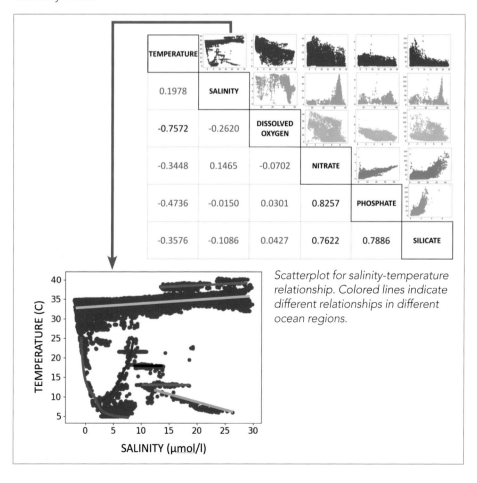

Scatterplot for salinity-temperature relationship. Colored lines indicate different relationships in different ocean regions.

GLOBAL EMERGING HOT SPOT RESULTS

Seagrass meadows are diminishing throughout the world because of climate change and human activity, including maritime traffic and pollution. Using emerging hot spot analysis to map changes in the suitability of potential seagrass environments indicates where these productive ecosystems might flourish and where they might not as oceans get warmer. Emerging hot spot analysis is based on statistical methods that identify patterns of change in hot spots over time. A hot spot has a statistically significant high value and is surrounded by other features with high values. Pinpointing the location of spatial clusters is important when scientists look for potential causes of the clustering, such as when and where a sudden die-off of seagrass might provide clues about the cause.

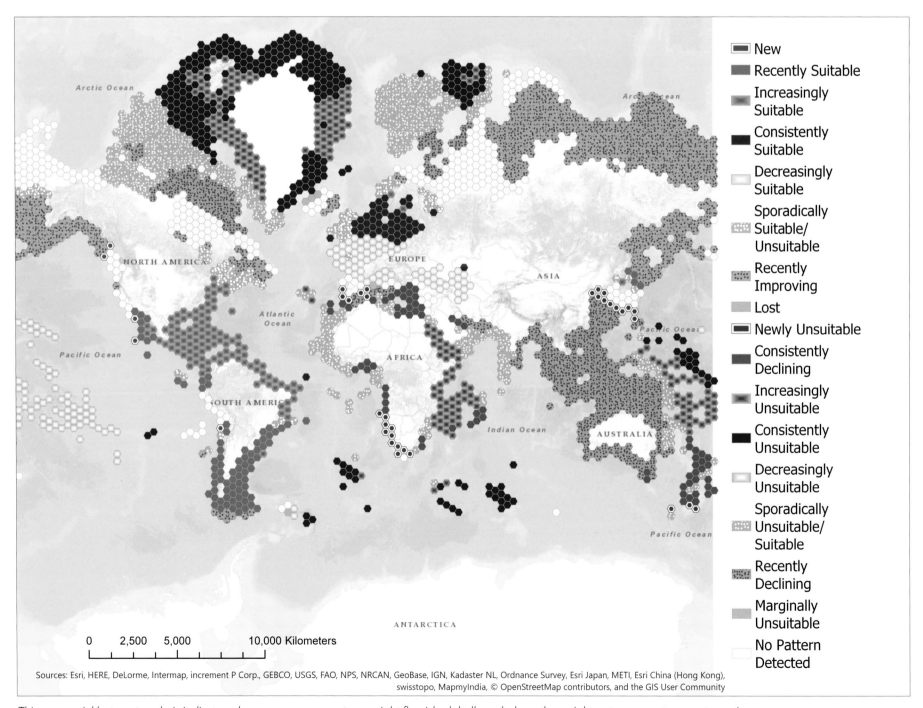

Legend:
- New
- Recently Suitable
- Increasingly Suitable
- Consistently Suitable
- Decreasingly Suitable
- Sporadically Suitable/Unsuitable
- Recently Improving
- Lost
- Newly Unsuitable
- Consistently Declining
- Increasingly Unsuitable
- Consistently Unsuitable
- Decreasingly Unsuitable
- Sporadically Unsuitable/Suitable
- Recently Declining
- Marginally Unsuitable
- No Pattern Detected

Sources: Esri, HERE, DeLorme, Intermap, increment P Corp., GEBCO, USGS, FAO, NPS, NRCAN, GeoBase, IGN, Kadaster NL, Ordnance Survey, Esri Japan, METI, Esri China (Hong Kong), swisstopo, MapmyIndia, © OpenStreetMap contributors, and the GIS User Community

This geospatial hot spot analysis indicates where seagrass ecosystems might flourish globally and where they might not as ocean temperatures rise.

PREDICTING FUTURE SEAGRASS HABITATS

We predicted seagrass habitats for each of the 20 years and combined the results into a spatio-temporal data structure called a *space-time cube*. Aggregating the data into a cube allowed us to use powerful GIS-based tools to discern patterns in space and time. For this analysis, we define a spatial pattern as clusters of high or low seagrass density, and a temporal pattern as increasing or decreasing intensity of that clustering. For each year of the analysis, if an area had higher than expected seagrass density we labeled it as suitable. Conversely, if an area had lower than expected seagrass density, we labeled it as unsuitable. The GIS method, emerging hot spot analysis, explores the patterns of suitable/nonsuitable across time and describes each area of the map with a label that summarizes the spatial and temporal patterns. For example, areas off the east coast of North America are increasingly suitable seagrass habitats across time. The map shows that Australia can potentially lose its seagrass habitat because of warming oceans and that places such as Siberia, which were not previously suitable seagrass habitats, will be suitable. Note that habitat suitability does not guarantee seagrass growth. Even if suitable habitat exists, seagrass species would somehow have to expand there.

Seagrass abundance spacetime cube displaying hot spot analysis results around the coast of Australia. The z-axis indicates the temperature increase (the lowest bin indicates current conditions, and the highest bin indicates a mean temperature increase of 2°C).

CONCLUSION

This chapter modeled future seagrass habitats in a data-driven workflow that used predictive machine-learning methods. We used random forest-based regression to model future ocean conditions under changing ocean temperatures and random forest-based classification to predict the presence or absence of seagrass. We mapped the results to illustrate the applicability of random forest to spatial problems and to compare modeling results to authoritative maps of seagrass distribution from the literature. The analysis of patterns across space and time found areas where seagrass will potentially be stressed in the future and areas where this vital species may become extinct. These spatiotemporal patterns show a mixed future for seagrass with significant habitat loss in some areas but potential habitat gain in others. These results show the potential destructive impacts of climate change on this valuable species.

ENDNOTES

1. R. J. Orth, T. J. B. Carruthers, W. C. Dennison, C. M. Duarte, J. W. Fourqurean, K. L., A. Heck, et al., "A Global Crisis for Seagrass Ecosystems," *BioScience* 56, no. 12, 1 (2006): 987–996, https://doi.org/10.1641/0006-3568(2006)56[987:AGCFSE]2.0.CO;2

2. P. L. Reynolds, E. Duffy, and N. Knowlton, "Seagrass and Seagrass Beds," *Smithsonian Ocean* (2016), https://ocean.si.edu/ocean-life/plants-algae/seagrass-and-seagrass-beds.

3. M. A. Hemminga and C. M. Duarte, *Seagrass Ecology* (Cambridge: Cambridge University Press, 2008).

4. Florida Fish and Wildlife Conservation Commission, Importance of Seagrass (September 11, 2018). Retrieved from http://myfwc.com/research/habitat/seagrasses/information/importance.

5. M. G. Grol, I. Nagelkerken, A. L. Rypel, and C. A. Layman, "Simple Ecological Trade-offs Give Rise to Emergent Cross-Ecosystem Distributions of a Coral Reef Fish," *Oecologia* 165, no. 1 (2011): 79–88.

6. C. M. Duarte, H. Kennedy, N. Marbà, and I. Hendriks, "Assessing the Capacity of Seagrass Meadows for Carbon Burial: Current Limitations and Future Strategies," *Ocean & Coastal Management* 83 (2013): 32–38.

7. M. Waycott, C. M. Duarte, T. J. Carruthers, R. J. Orth, W. C. Dennison, S. Olyarnik, et al., "Accelerating Loss of Seagrasses across the Globe Threatens Coastal Ecosystems," *Proceedings of the National Academy of Sciences* 106, no. 30 (2009): 12377–12381.

8. F. T. Mackenzie, L. M. Ver, and A. Lerman, "Coastal-Zone Biogeochemical Dynamics under Global Warming," *International Geology Review* 42, no. 3 (2000): 193–206, https://doi.org/10.1080/00206810009465077.

9. J. C. Duque, R. Ramos, and J. Surinach, "Supervised Regionalization Methods: A Survey," *International Regional Science Review* 30, no. 3 (2007): 195–220, https://doi.org/10.1177/0160017607301605.

10. R. M. Assunção, M. C. Neves, G. Câmara, and C. Da Costa Freitas, "Efficient Regionalization Techniques for Socio-economic Geographical Units Using Minimum Spanning Trees," *International Journal of Geographical Information Science* 20, no. 7 (2006): 797–811, https://doi.org/10.1080/13658810600665111.

11. T. Calinski and J. Harabasz, "A Dendrite Method for Cluster Analysis," *Communications in Statistics—Theory and Methods* 3, no. 1 (1974): 1–27, https://doi.org/10.1080/03610927408827101.

12. R. J. Orth, T. J. B. Carruthers, W. C. Dennison, C. M. Duarte, J. W. Fourqurean, K. L., A. Heck, et al., "A Global Crisis for Seagrass Ecosystems," *BioScience* 56, no. 12, 1 (2006): 987–996, https://doi.org/10.1641/0006-3568(2006)56[987:AGCFSE]2.0.CO;2

13. M. Meinshausen, N. Meinshausen, W. Hare, S. C. B. Raper, K. Frieler, R. Knutti, et al., "Greenhouse-Gas Emission Targets for Limiting Global Warming to 2°C," *Nature* 458, no. 7242 (2009): 1158–1162, https://doi.org/10.1038/nature08017.

14. D. Wright, R. Sayre, S. Breyer, K. Butler, K. VanGraafeiland, K. Goodin, et al., "The Ecological Marine Units Project as a Framework for Collaborative Data Exploration, Distribution, and Knowledge Building," 19th EGU General Assembly, EGU2017, Proceedings from the Conference, held 23–28 April, 2017, in Vienna, Austria, 6051, 19, 6051, retrieved from http://adsabs.harvard.edu/abs/2017EGUGA..19.6051W.

Pipefish and tube snouts photo by Rhinopias, licensed under Creative Commons Attribution-Share Alike 4.0 International.

Australian giant cuttlefish photo on page 60 by Richard Ling @ www.rling.com.

Green Sea Turtle photo on page 61 by P. Lindgren, licensed under Creative Commons Attribution-Share Alike 4.0 International.

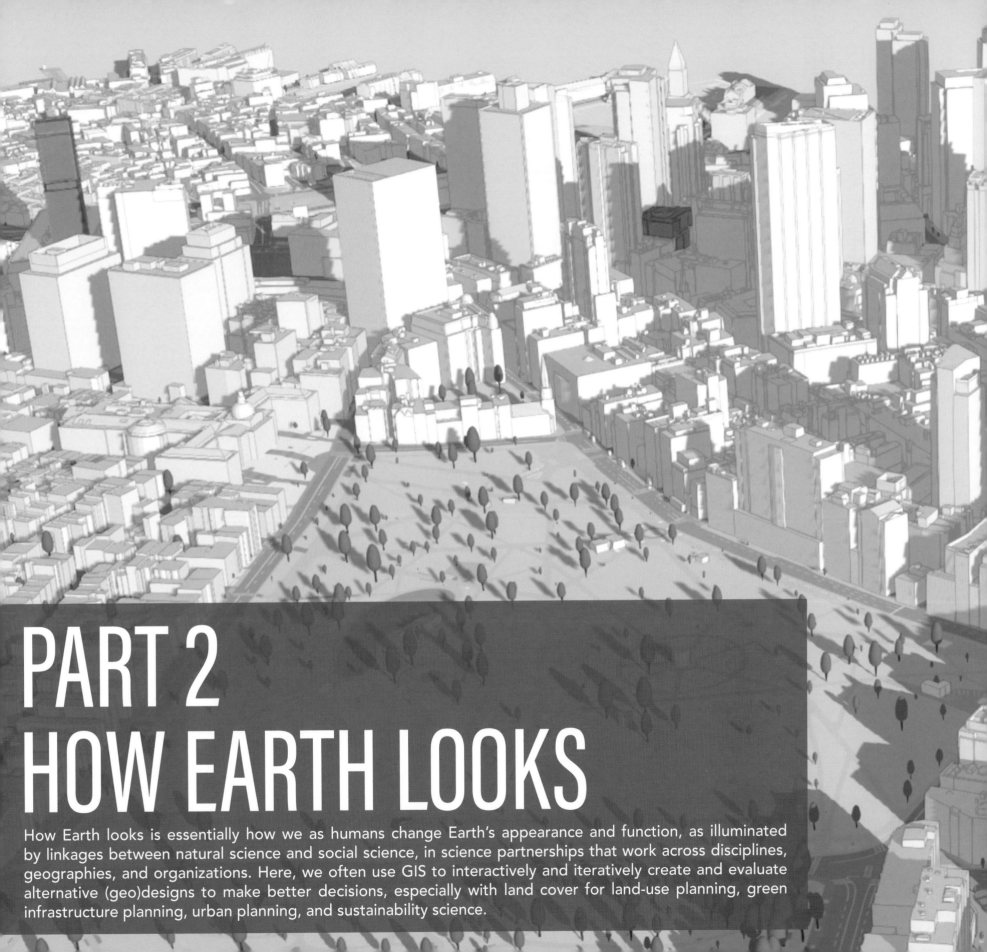

PART 2
HOW EARTH LOOKS

How Earth looks is essentially how we as humans change Earth's appearance and function, as illuminated by linkages between natural science and social science, in science partnerships that work across disciplines, geographies, and organizations. Here, we often use GIS to interactively and iteratively create and evaluate alternative (geo)designs to make better decisions, especially with land cover for land-use planning, green infrastructure planning, urban planning, and sustainability science.

This GIS view of downtown Boston simulates shadow patterns at 1:40 PM on January 9, 2019. Try the interactive web app created by the Office of GIS, Boston Planning and Development Agency linked at GISforScience.com.

EXTREME HEAT EVENTS IN A CHANGING CLIMATE

Extreme heat is a major public health concern, and in response, scientists are using GIS to aid public officials in monitoring the frequency and intensity of forthcoming extreme heat events.

By Olga Wilhelmi and Jennifer Boehnert, National Center for Atmospheric Research (NCAR)

Extreme summer heat is felt all the more intensely in urban areas.

EXTREME HEAT AND HUMAN HEALTH

Extreme heat is a major public health concern. Hot temperatures combined with high humidity cause human discomfort and may lead to serious illness and death. Impacts of heat stress on human health have been observed around the world— across different geographies and levels of economic development. In the United States, extreme heat caused more deaths than any other weather-related hazard during the 30-year-period from 1988 to 2017, according to the National Weather Service (NWS). The US Centers for Disease Control and Prevention (CDC) estimates an average of 600 people die every year from exposure to extreme heat (CDC 2012).

Urban residents are especially vulnerable to heat (Wilhelmi et al. 2012). High concentration of human activities (e.g., heat emissions from cars, buildings, and industry) and features of urban environment (e.g., buildings and paved impervious surfaces) create an urban heat island (UHI) effect. UHI is a phenomenon that characterizes local-scale temperature differences within cities (e.g., the difference in temperature between parks and paved parking lots) and between cities and surrounding nonurban areas (Oleson et al. 2015). UHI generally contributes to an increased heat exposure; however, the intensity of this effect depends on factors such as city size and population density, surface characteristics, geographic location, and regional climate.

Negative health impacts of extreme heat, such as heat rash, exhaustion, and stroke, can occur even during normal hot summer months because of prolonged exposure to heat and an inability to cool off. However, higher numbers of heat-related health impacts are typically observed during a heat wave, when daily temperatures exceed a range normal for a given climate and the local setting (Anderson and Bell 2010). Local climate characteristics, UHI effect, and a combination of temperature, humidity, wind speed, and solar radiation often determine whether any given location experiences an extreme heat event or a heat wave (Morss et al. 2011; Wilhelmi et al. 2012). Weather conditions, population and housing characteristics, and the readiness of local government to mitigate heat hazards often determine the magnitude of heat-related illnesses and deaths (Wilhelmi et al. 2012).

Future urbanization, demographic trends, and climate change suggest that extreme heat will remain a major hazard in the United States (Field et al. 2012; Oleson et al. 2015), disproportionately affecting the health of vulnerable populations. Moreover, future exposure of the US population to extreme heat is projected to increase. As global warming continues, extreme heat events are projected to increase in severity, frequency, duration, and spatial extent (Meehl and Tebaldi 2004; Oleson et al. 2015). As cities and counties begin to adapt to the changing climate and plan for future extreme events, it is imperative to consider a range of strategies for effective heat preparedness and response, including potential changes to heat warnings and advisories.

This chapter describes heat warning and response practices in the United States, discusses the impact of climate change on extreme heat, and explores how frequency and intensity of extreme heat events may change by the middle of the twenty-first century. Using an example from the Houston-Galveston area in Texas, we present a GIS-based methodology for climate data analysis and visualization and demonstrate a workflow in ArcGIS® Pro software.

MEASURING HEAT AND WARNING POPULATION ABOUT HEAT STRESS

Extreme heat events and their effects on human health can be challenging to define or measure. Heat waves have no standard definition (Robinson 2001), and efforts to develop universal heat stress measures have continued for several decades (Epstein and Moran 2006). Even though heat waves are meteorological phenomena, many definitions attempt to represent the interaction between the thermal environment and human health (Meehl and Tebaldi 2004; Robinson 2001). Efforts have been made to estimate heat stress from local climate characteristics, human physiology, and work conditions, and to combine different meteorological and physiological attributes into a heat-stress index (Epstein and Moran 2006).

Heat indices are created because ambient temperature alone may not accurately represent human discomfort from heat, especially when humidity is high. Most of heat indices are calculated using various measures of temperature and humidity (e.g., relative humidity, dew point), and some include solar radiation and wind speed (Epstein and Moran 2006). Certain heat indices are created for a specific user group or geography, while others are used for more general purposes, but all aim to estimate human thermal comfort/discomfort and safety limits.

In the United States, the National Oceanic and Atmospheric Administration (NOAA) National Weather Service (NWS) has been using a heat index (HI) in its operational weather forecasting. The NWS heat index is calculated from ambient temperature and relative humidity (NOAA 2014). NWS forecast offices across the country issue heat watches, advisories, and excessive heat warnings when the HI exceeds thresholds considered dangerous even for healthy individuals. The NWS developed its excessive heat criteria in collaboration with local weather forecasters, emergency managers, and researchers across the United States. While general-purpose classification of the HI exists, the specific heat thresholds required for the issuance of heat advisories and warnings are place-specific and vary among NWS weather forecast offices (WFOs).

The map here shows locations of the WFOs and their respective county warning areas (CWA)—the group of counties for which WFOs are responsible for issuing weather warnings. A total of 122 WFOs span the continental United States, Alaska, Hawaii, Guam, and Puerto Rico. WFOs provide weather services to the counties in their respective CWAs. These services include issuing excessive heat watches (when conditions favorable for an excessive heat event meet or exceed local heat warning criteria in the next 24 to 72 hours), heat advisories (when HI values are forecast to meet or exceed local heat advisory criteria within the next 12 to 24 hours), and heat warnings (when the HI is expected to reach or exceed locally determined thresholds within the next 12 to 24 hours). Heat advisory values typically range from 100°F in the northern states to 105°F in the southern states. Similarly, heat warning thresholds generally range from more than 105°F in the north to more than 110°F in the south. When nighttime lows are considered in heat advisory or warning criteria, forecasters often use a nighttime temperature of 75°F as a typical threshold, with some variability in temperature range between northern and southern states (Hawkins et al. 2016).

Although these general guidelines exist, the heat watch, advisory, and warning criteria can vary even between locations that have close geographic proximity and depend on local climate and the impact of extreme heat on local populations (Hawkins et al. 2016). Some WFOs use only daily maximum HI values, others use a combination of HI and daily minimum temperature values, and some WFOs include a time span in their excessive heat criteria (e.g., heat index values must meet or exceed a certain threshold for at least two hours or consecutive days). For example, the WFO for Baltimore/Washington, DC, issues an excessive heat warning when the HI is expected to reach or exceed 110°F. The WFO in New Orleans/Baton Rouge issues excessive heat warnings when the HI is forecast to exceed 113°F or if the daily maximum temperature reaches or exceeds 105°F .

Locations of WFOs and their respective county warning areas (CWA)—the group of counties for which WFOs are responsible for issuing weather warnings in the continental US. Data courtesy of NOAA NWS. (Alaska, Hawaii, Guam, and Puerto Rico omitted.)

RESPONDING TO EXTREME HEAT

Heat advisories and warnings intend to raise public awareness and help mitigate negative impacts of heat on human health. The NWS heat advisories and warnings serve as signals for the public to take protective measures, such as those outlined by the CDC for the elderly, very young children, outdoor workers, and athletes (CDC 2018).

In many cities, heat alerts also trigger coordinated emergency operations, civil protection, and public health interventions (Bassil and Cole 2010). For example, a heat warning may alert hospitals to prepare for more emergency calls. Other examples may include media announcements, opening cooling centers, activating programs that work with vulnerable groups (e.g., elderly, homebound, homeless), and changing activity levels among outdoor workers and athletes. In some cases, various agencies must coordinate to reposition resources and activate programs to minimize health risks. Comprehensive heat response requires involvement of many city departments and nongovernmental organizations (Bernard and McGeehin 2004).

Because access to cooling helps reduce heat-health risks, many cities developed programs that designate public air-conditioned buildings as cooling centers or heat refuges. As shown on the map identifying cooling centers in Houston, public libraries, community centers, and other city buildings can serve as cooling centers by opening their doors to the public during normal business hours or by extending hours of operation; for example, if an extreme heat event occurs during the weekend.

In addition to providing centralized cooling centers, several cities have implemented targeted, neighborhood-based programs to reach the most at-risk populations, such as seniors and people with disabilities. For example, the Cool Neighborhood NYC program aims to reduce heat-related health impacts, not only by reducing UHI but also by strengthening social networks and providing climate-risk training for home health aides. In other cases, partnership with local utility companies ensures that they cannot shut off residential water or electricity services during a heat wave. Heat warnings may also trigger changes to regulations such as evictions and outside work requirements. Overall, comprehensive heat response plans are necessary to reduce heat-related morbidity and mortality, and require coordinated actions across government authorities, the private sector, and the public. Heat warnings play a major role in setting these plans in motion.

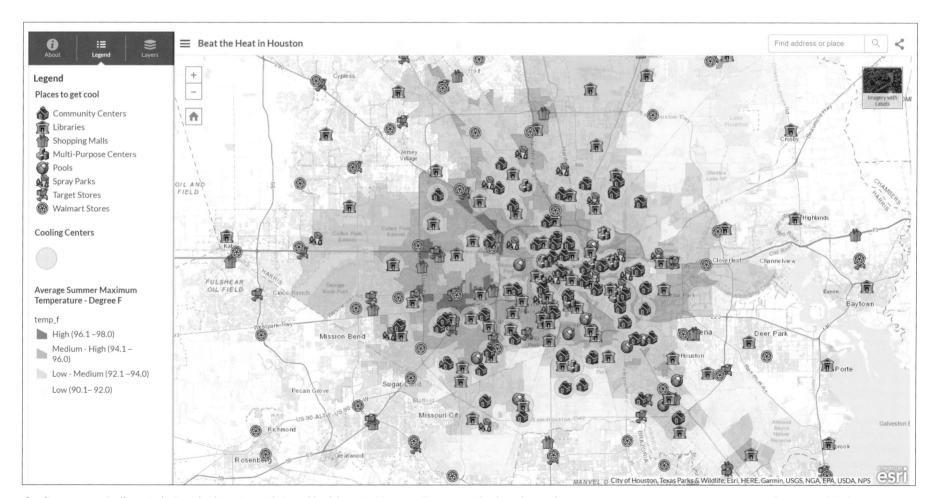

Cooling centers (yellow circles) and other air-conditioned buildings in Houston, Texas, are displayed over the average maximum summer temperature by census block group. Beat the Heat in Houston interactive map is created using ArcGIS Online to raise public awareness about heat hazards and inform the public about ways to protect oneself from heat. https://toolkit.climate.gov/tool/beat-heat-houston.

EXTREME HEAT IN A CHANGING CLIMATE

Climate change exacerbates the impacts of extreme heat by increasing the number and severity of naturally occurring rare events, like heat waves. Even small increases of globally averaged temperatures (shown on a map here) can result in a relatively large increase in temperature extremes (Morss et al. 2011). As climate continues to warm, we can expect more severe heat events (hotter days and nights) and frequency (number of hot days and nights), and longer duration (prolonged heat events) (IPCC 2012). Recent reports (e.g., Fourth National Climate Assessment Report; Reidmiller et al. 2018) indicate that many US counties and cities will have more dangerously hot days and nights by the middle of the twenty-first century. Intensity of heat waves in cities is projected to be greater than in rural areas because of the UHI effect (Stone et al. 2012). Climate scientists examine changes in extreme heat (or heat stress days and nights) using absolute threshold measures (e.g., when daily maximum temperature exceeds 100 degrees) or relative measures (e.g., when temperature exceeds the ninetieth percentile of a climatological data record for a given location) (Morss et al. 2011). Numerical climate modeling helps examine extreme heat events in the context of changing climate, using one or more scenarios of future increased greenhouse gases (Oleson et al. 2015). The resulting model output can then be analyzed for changes in future extreme events using the selected thresholds for an area of interest.

Researchers commonly assess extreme heat using projections of air temperature from climate models (Meehl and Tebaldi 2004). But since humidity can aggravate the physiological effects of high temperature (Epstein and Moran 2006), recent efforts have tried to simulate future heat indices (Oleson et al. 2015). Given that the NWS heat index is widely used in the United States, exploring potential changes in the days that meet or exceed the current heat warning or advisory criteria could help local governments prepare for future extreme heat events.

GIS can support the analysis of future climate scenarios and provide effective visualization and decision-support platforms for researchers and decision-makers (Wilhelmi et al. 2015). We will demonstrate a GIS workflow for exploring projected changes in extreme heat in the next section.

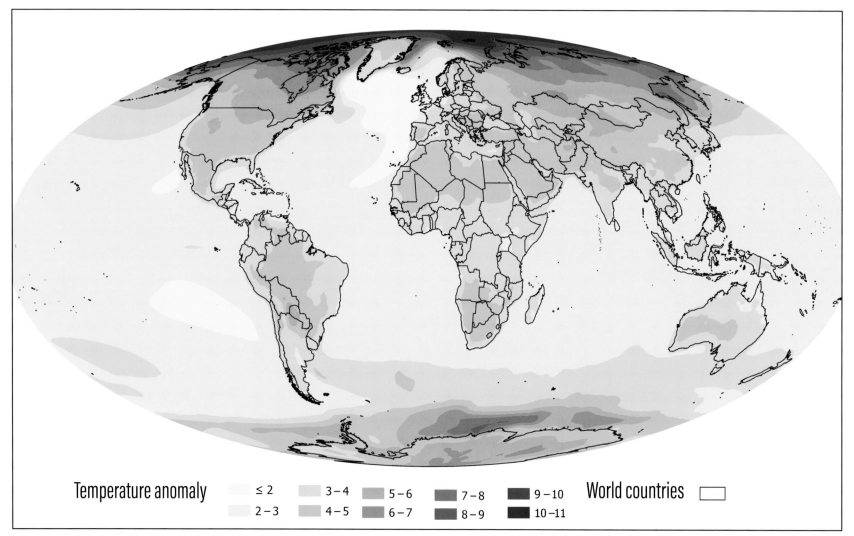

Temperature anomaly ≤ 2 | 2–3 | 3–4 | 4–5 | 5–6 | 6–7 | 7–8 | 8–9 | 9–10 | 10–11   **World countries**

Global temperature anomaly map for RCP 8.5 for the end of the century (2081–2100) compared with present day (1986–2005). Data source: CCSM4, NCAR.

MODELING AND MAPPING SUMMER HEAT IN THE UNITED STATES

To map changes in extreme heat across the United States, we used projections from the NCAR's Community Land Model Urban (CLMU) (Oleson et al. 2015). CLMU is a numerical model designed to simulate the climate of urban environments within a global climate model, particularly the temperature where people live. This model allows investigation of how climate change affects the UHI. Oleson et al. (2015) used the CLMU to quantify differences between rural and urban summer heat over the continental United States and southern Canada for present-day climate (1986–2005) and for the middle of the twenty-first century (2046–2065). The study used the "business as usual" climate change scenario (i.e., Representative Concentration Pathway 8.5; Moss et al. 2008). Spatial resolution of the model simulations was one-eighth degree in latitude and longitude or approximately 10 square kilometers. The results of the model simulations included hourly data for air temperature and HI for the summer season (May–September) for both present-day and future climates. (1986–2005 is considered "present day" for climate scientists.)

To examine heat extremes, we first used CLMU model output to calculate the daily maximum temperature and daily maximum HI values from hourly inputs for further analysis in ArcGIS Pro. Because climate models generate data in the netCDF (network Common Data Form; https://www.unidata.ucar.edu/software/netcdf/) format, we performed the initial data processing using an open source software, Climate Data Operators (CDO; https://code.mpimet.mpg.de/projects/cdo/embedded/cdo.pdf). CDO is designed to work with large climate datasets and performs statistical functions efficiently.

Estimating potential change in summertime heat can be done by calculating average meteorological values during a long period, for example, 20 years. The 20-year averages "smooth out" single-year extreme events and are useful for analyzing long-term climate trends. To compare how the summer heat conditions may be different by 2050, we used the daily maximum HI data to create a 20-year summer average maximum HI for the present-day climate (1986–2005) and for the mid-century (2046–2065) climate. By calculating long-term average conditions, we "smoothed out" the most extreme HI values and analyzed the average high heat conditions prevalent during the boreal summer season (May through September). By comparing the two HI maps, we can observe generally hotter conditions across the United States during the mid-century.

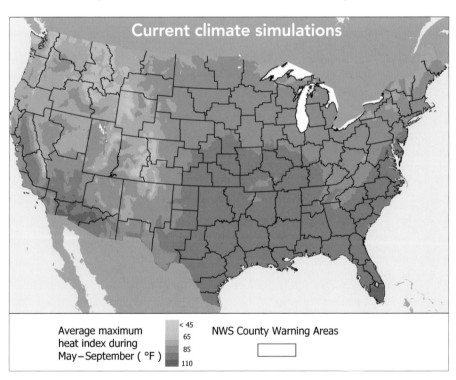

Current climate (1986–2005) climate summer (May–September) average maximum heat index, simulated by CLMU (Oleson et al. 2015).

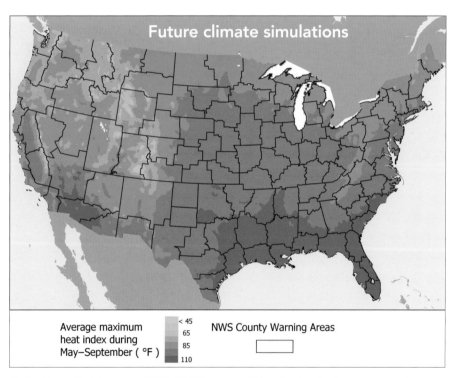

Projections of mid-century (2046–2065) summer (May–September) average maximum heat index, simulated by CLMU (Oleson et al. 2015).

VISUALIZING SPATIAL PATTERNS OF CHANGE

Anomaly in climate science is a deviation of a meteorological variable (e.g., temperature or HI) from the normal (mean) value. Here, we use climate from 1986 to 2005 as a norm. We can calculate the anomaly by comparing (subtracting) the long-term average of model simulations from an average of future climate simulations. To visualize the difference between average summer heat in present and future climates, we used the ArcGIS tool *raster calculator* to create a maximum HI anomaly map. It shows greater heat anomalies (up to 10 degrees) in the South and Southeast of the United States. The Midwest region and the mountainous areas across the western United States are also projected to have more intense summer

heat by 2050, using the representative concentration pathway (RCP) 8.5, business-as-usual climate scenario.

Climate anomalies help us visualize general climate trends and assess the degree of change. However, as mentioned earlier, when we calculate long-term averages, we smooth out some of the most extreme heat events. To better understand the interannual variability in extreme heat and estimate potential change in the intensity and frequency of the summer heat waves, we can analyze mulitidimensional data in ArcGIS Pro.

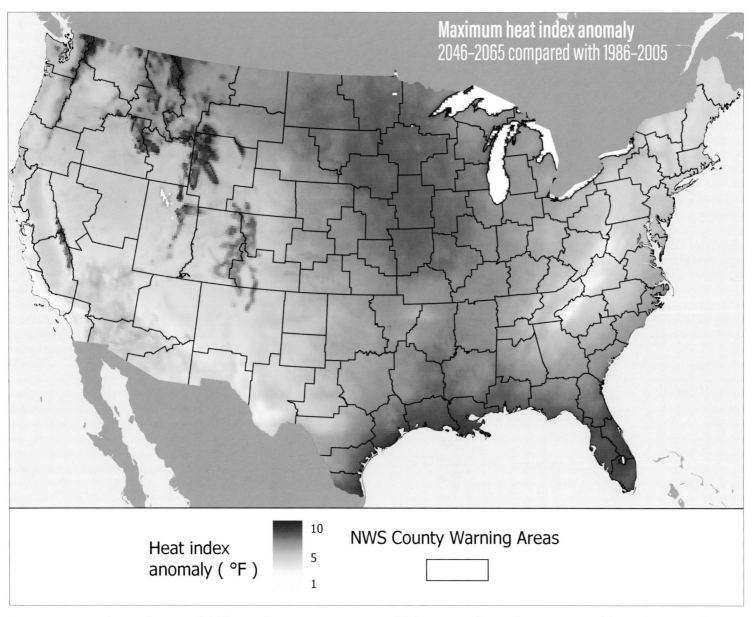

Average maximum heat index anomaly (difference between summer mean of daily maximum heat index in present and future climates) in the continental US (simulated by CLMU; Oleson et al. 2015).

ANALYZING SPATIOTEMPORAL CHANGES OF EXTREME HEAT: A HOUSTON EXAMPLE

Residents of southern Texas have a lot of experience with extreme heat. The summer season, from late May through September, is characterized by hot weather conditions. Humid, subtropical climate of the gulf coastal plain has summer temperatures often above 90°F. The NWS's Houston-Galveston WFO, located in Dickinson, Texas, provides weather warnings for Houston-Galveston CWA. The Houston-Galveston CWA comprises 23 counties and includes several large metropolitan areas, including Houston, Galveston, and College Station (see map on page 75). Houston, the largest city in Texas and the fourth largest in the United States, has a fast-growing, multicultural population of 2.3 million. The map below illustrates the role of UHI effect in the distribution of summer heat across Harris County. It shows summer average nighttime temperature, simulated using the High Resolution Land Data Assimilation System (HRLDAS) and aggregated to the census block groups.

The Houston-Galveston WFO issues heat advisory when the daily maximum heat index (HI max) reaches or exceeds 108°F or when the daily maximum temperature (T max) reaches or exceeds 103°F. The excessive heat warnings are issued when the HI reaches or exceeds 113°F or T max is equal to or greater than 105°F. Heat advisories and warnings can be issued for the entire CWA or for the select counties. Historically, on average, the Houston-Galveston WFO issues three heat advisories or warnings during summer season (City of Houston 2018). To investigate potential changes in the frequency and intensity of extreme heat events in the Houston-Galveston CWA, we used these heat thresholds and the climate model simulations from Oleson et al. (2015).

The climate model output from Oleson et al. (2015) is stored in the netCDF data format. The NetCDF data format is commonly used in the atmospheric sciences because it stores data as multidimensional arrays. Spatial dimensions such as latitude and longitude are common for most GIS data; however, climate data also can have temporal and pressure dimensions. The data used in our study (i.e., HI max and T max) have three dimensions: latitude, longitude, and time. In the next section we present a workflow to analyze spatiotemporal changes of extreme heat using these multidimensional data. Our data processing and analysis workflow for the Houston-Galveston CWA can be applied to any other NWS WFOs and their corresponding CWAs.

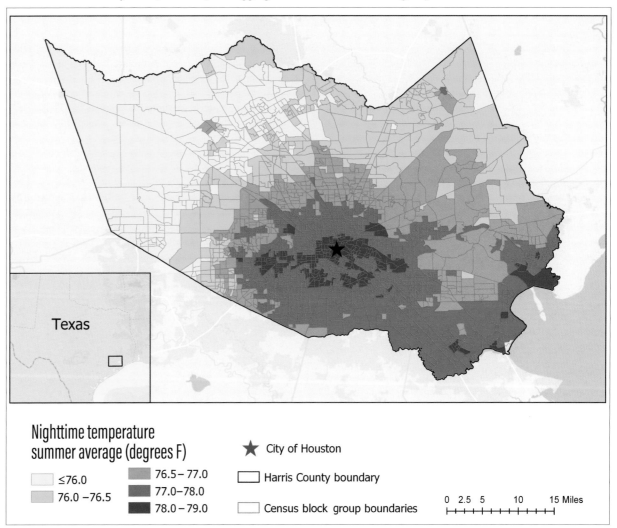

Nighttime temperature summer average (degrees F)

- ≤76.0
- 76.0 – 76.5
- 76.5 – 77.0
- 77.0 – 78.0
- 78.0 – 79.0

★ City of Houston

▢ Harris County boundary

▢ Census block group boundaries

0 2.5 5 10 15 Miles

Nighttime average summer temperature in Harris County Texas illustrates the UHI and local climate characteristics. The meteorology of Houston's urban heat island was simulated using the High Resolution Land Data Assimilation System.

Workflow

This ModelBuilder workflow in ArcGIS Pro shows the spatiotemporal analysis of extreme heat events. We used the same workflow with both present-day and future climate model simulations for the HI and the maximum temperature variables. The result of the workflow is a table that shows the daily maximum HI and temperature for each day from all input points for the 20-year period that exceed excessive heat alert criteria.

These steps describe the process in greater detail.

Step 1: Read netCDF data as feature and raster layers

ArcGIS Pro can read netCDF format as points, rasters, or tables using the Multidimensional toolbox. In the first step, we brought the netCDF data (daily HI max and T max) from the two model simulations into ArcGIS Pro as points and rasters (i.e., using the Multidimensional Feature Layer and Make Multidimensional Raster Layer functions). In this step, any period can be selected for a temporal dimension. We used the points as positional reference for further analysis of this multidimensional raster dataset (step 3).

Step 2: Join climate data with area of interest

In the second step, we created a spatial join between the climate model point file and the Houston-Galveston CWA polygon shapefile they fall within. The output feature class contains climate data points and all the fields from the NWS CWA shapefile. Once the points have the CWA information, we can select by attribute to choose only the points that fall within the Houston-Galveston WFO CWA.

Step 3: Sample multidimensional data to create time series

The Sample analysis tool creates a table representing data from the multidimensional raster. The output table contains all the values from all the times for all the grid cells from the netCDF raster data created in step 1. In the workflow seen here, the green ellipse called "Time series table" represents an output from the Sample tool and contains all daily HI values for the summer months for 20 years for each grid cell that fell within the Houston–Galveston WFO CWA. We used this large table, containing 963,900 rows, in the next step to query the maximum HI and temperature values from all grid cells for each day to identify the daily maximum heat for our area of interest.

Step 4: Generate table of threshold exceedances

To select the highest values from the output table generated through the Sample tool, we ran a Summary Statistics tool. Using Maximum as the statistics type and Time as the case field, we generated a table that contained daily maximum HI and temperature. We can use this table for further analysis of the frequency, intensity, and duration of extreme heat events and for visualizing the results using ArcGIS Pro data visualization and charting tools.

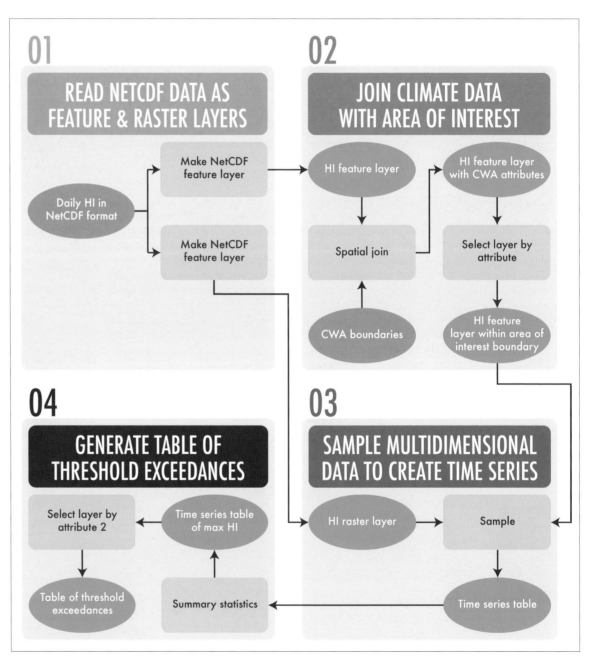

Workflow using Model Builder to generate a table with daily HI max that exceed heat advisory thresholds for a particular CWA.

VISUALIZING TEMPORAL CHANGES IN EXTREME HEAT EVENTS

Data analysis showed that the Houston-Galveston CWA may see a nearly six-degree increase in the summer average daily maximum HI by 2050, with a possible tenfold increase in the number of extreme heat days. Present-day climate model simulations show on average three extreme heat events per summer that would fall within the current category for heat advisory. By 2050, using the RCP 8.5 scenario, as many as 30 days during summer months may warrant a heat advisory or an excessive heat warning. Heat events may also be longer-lasting and more severe. Under current climate simulations, 13 percent of all heat alerts would classify as excessive heat warnings (with a HImax ranging between 135°F and 120°F). By the middle of the twenty-first century, excessive heat warnings would increase to 32 percent of all heat alerts. The severity of heat also would increase, with the daily maximum HI increasing to 136°F.

The calendar grids seen here on the top of the page were generated using ArcGIS Pro. The daily data was organized in a table where years were the rows and days were the columns. When we organize the data this way, we can save it as an Esri ASCII raster format. We can then use this format to transfer information to or from other cell-based raster systems. It contains header information that defines the properties of the raster such as the cell size, the number of rows and columns, and the coordinates of the origin of the raster. The data information is specified in space-delimited row-major order, with each row separated by a carriage return. This format allows us to use days and years as x,y coordinates and render the data as a grid on the map, thus creating a multidimensional temporal data visualization.

The circular diagrams another innovative visualization of the projected changes in the frequency of extreme heat events throughout the summer season in the Houston-Galveston CWA. The data clock—a new charting capability in ArcGIS Pro—displays the frequency of excessive heat advisories and warnings instead of daily HI values as shown on the calendar grid. Using data clocks, we can observe the expansion of the extreme heat occurrences into early and late summer. This chart type is also useful to visualize the number of times the heat advisory and warning criteria has been met each month during the respective 20-year periods. Both visualizations help communicate the increasing heat stress risk for the populations of the 23 counties in the Houston-Galveston region. The results also indicate the need to prepare for future extreme heat in the urban centers and rural communities.

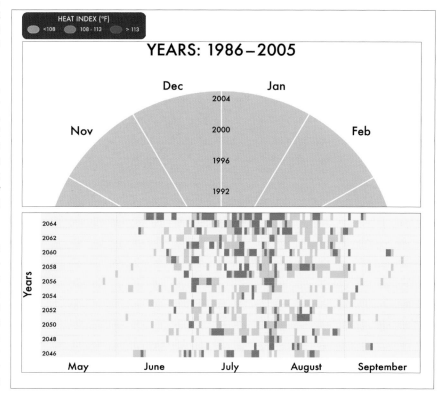

Occurrence of excessive heat advisories and warnings in present day climate (top) and under future climate scenario (bottom) [RCP 8.5] in 2050.

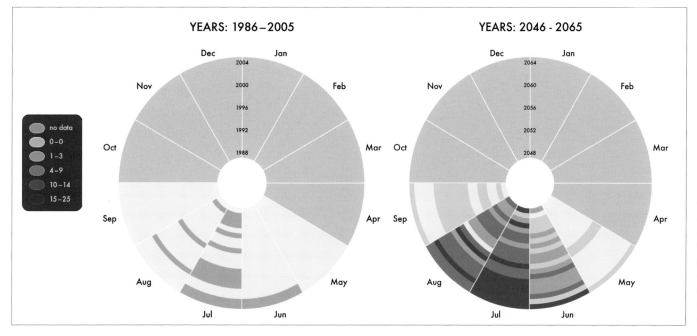

The 1986–2005 data clock (far left) shows that the months of May, June, and September for those years have no or few occurrences of excessive heat episodes, with most heat waves taking place in July and August. The 2046–2065 clock shows a general expansion of the extreme heat season and a higher frequency of excessive heat alerts throughout June, July, and August.

CONCLUSION

This chapter discussed the relationships between extreme heat, human health, and climate change. As a major public health concern, extreme heat is projected to increase in frequency, intensity, and duration toward the middle of the twenty-first century. US cities and counties have developed locally relevant place-based criteria for warning their populations about dangers of extreme heat. Because of the changing climate, rapid urbanization, and growing number of heat-sensitive populations, extreme heat in some parts of the United States, such as southern Texas, may become a new norm rather than a rare event during the summer season. This chapter demonstrated how to analyze and visualize potential changes in extreme heat using climate model simulations and ArcGIS Pro. We illustrated spatial patterns of changing heat across the country and explored temporal changes in extreme heat events for our area of interest.

Using an example of southern Texas—NWS Houston-Galveston CWA—we applied the location-specific excessive heat advisory and warning criteria to model simulated heat index and air temperature in current and future climates to explore potential changes in extreme heat events. We developed an ArcGIS Pro workflow that can be applied in other CWAs across the United States to better understand how climate change may affect frequency, intensity, and duration of heat waves. This workflow also can be used to answer other research questions that rely on multidimensional data and a specific geographic boundary (e.g., country boundary or watershed).

We used the operational heat alert criteria in our analysis with the understanding that heat warnings set in motion coordinated emergency operations across public and private sectors. It is important to consider the implications of changing heat extremes because they span a wide range of impacts on population, infrastructure, and emergency response. Developing public-private partnerships and conducting future research on the interactions between heat and health and on human adaptation to changing extreme events can support stakeholders as they develop heat preparedness and response strategies. From this work, we can better understand future exposure to extreme heat and facilitate the planning and adaptation process. Our work also showed that GIS effectively facilitates data analysis and science communication.

ENDNOTES

Anderson, G. B., and M. I. Bell. 2010. "Heat Waves in the United States: Mortality Risk during Heat Waves and Effect Modification by Heat Wave Characteristics in 43 US Communities. " *Environmental Health Perspectives* 119, no. 2: 210–218.

Bassil, K., and D. Cole. 2010. "Effectiveness of Public Health Interventions in Reducing Morbidity and Mortality during Heat Episodes: A Structured Review." *International Journal of Environmental Research and Public Health* 7, no. 3: 991–1001.

Bernard, S. M., and M. A. McGeehin. 2004. "Municipal Heat Wave Response Plans." *American Journal of Public Health* 94, no. 9: 1520–1522.

City of Houston. 2018. Hazard Mitigation Action Plan. Retrieved from https://s3-us-west-2.amazonaws.com/uasi-jtti/wp-content/uploads/sites/29/2018/05/16202024/City-of-Houston-HMAP-Update-3.13.18-3.12.23-PUBLIC-COPY_compressed.pdf.

Centers for Disease Control and Prevention (CDC). 2012. "QuickStats: Number of Heat-Related Deaths, by Sex—National Vital Statistics System, United States, 1999–2010." *Morbidity and Mortality Weekly Report (MMWR)* 61, no. 36: 729.

CDC. 2018. Extreme Heat. Retrieved from https://www.cdc.gov/disasters/extremeheat/index.html.

Epstein, Y., and D. S. Moran. 2006. "Thermal Comfort and the Heat Stress Indices." *Industrial Health* 44, no. 3: 388–398.

Field, C. B., V. Barros, T. F. Stocker, and Q. Dahe, eds. 2012. *Managing the Risks of Extreme Events and Disasters to Advance Climate Change Adaptation: Special Report of the Intergovernmental Panel on Climate Change*. Cambridge: Cambridge University Press.

Hawkins, M. D., V. Brown, and J. Ferrell. 2017. "Assessment of NOAA National Weather Service Methods to Warn for Extreme Heat Events." *Weather, Climate, and Society* 9, no. 1: 5–13.

Meehl, G. A., and C. Tebaldi. 2004. "More Intense, More Frequent, and Longer Lasting Heat Waves in the 21st Century." *Science* 305, no. 5686: 994±7. https://doi.org/10.1126/science.1098704.

Morss, R. E., O. V. Wilhelmi, G. A. Meehl, and L. Dilling. 2011. "Improving Societal Outcomes of Extreme Weather in a Changing Climate: An Integrated Perspective." *Annual Review of Environment and Resources* 36: 1–25.

Moss, R., M. Babiker, S. Brinkman, E. Calvo, T. Carter, J. Edmonds, et al. 2008. "Towards New Scenarios for Analysis of Emissions, Climate Change, Impacts, and Response Strategies." Geneva: Intergovernmental Panel on Climate Change: 132.

NOAA National Weather Service. 2014. The Heat Index Equation. Retrieved from https://www.wpc.ncep.noaa.gov/html/heatindex_equation.shtml.

Oleson, K. W., A. Monaghan, O. Wilhelmi, M. Barlage, N. Brunsell, J. Feddema, L. Hu, and D. F. Steinhoff. 2015. "Interactions between Urbanization, Heat stress, and Climate Change." *Climatic Change* 129: 525–541. doi:10.1007/s10584-013-0936-8.

Reidmiller, D. R., C. W. Avery, D. R. Easterling, K. E. Kunkel, K. L. M. Lewis, T. K. Maycock, and B. C. Stewart, eds. 2018. *Fourth National Climate Assessment Volume II: Impacts, Risks, and Adaptation in the United States, Report in Brief*. US Global Change Research Program, Washington, DC: US Government Publishing Office.

Robinson, P. J. 2001. "On the Definition of a Heat Wave." *Journal of Applied Meteorology* 40, no. 4: 762–775.

Stone, B. 2012. *The City and the Coming Climate: Climate Change in the Places We Live*. New York: Cambridge University Press.

Wilhelmi, O., A. de Sherbinin, and M. Hayden. 2012. "Exposure to Heat Stress in Urban Environments: Current Status and Future Prospects in a Changing Climate." *In Ecologies and Politics of Health*, edited by B. King and K. Crews, chapter 12. London: Routledge.

Wilhelmi, O., K. Sampson, and J. Boehnert. 2015. "Exploring Future Climates in a GIS." In *Mapping and Modeling Weather and Climate with GIS*, edited by Lori Armstrong, et al. Redlands: Esri Press.

FINDING A WAY HOME

This chapter presents a glimpse into the homelessness crisis taking place across America and describes how GIS can help cities, agencies, and spatial analysts understand, prevent, and manage this human dilemma.

By Lauren Griffin and Este Geraghty, Esri

A block-long tent encampment on a sidewalk in Los Angeles. As with a lot of places in America, the homelessness crisis is in plain sight. (Photo by Russ Allison Loar.)

HOMELESSNESS ACROSS AMERICA

When did homelessness become such an accepted and entrenched part of the urban landscape? It seems we barely notice as we drive past the homeless man lying in the street or the homeless woman holding a handwritten appeal for help.

But now in America, we are finding people experiencing homelessness in every state, in communities large and small, and in places where we never would have imagined encountering this growing national crisis.

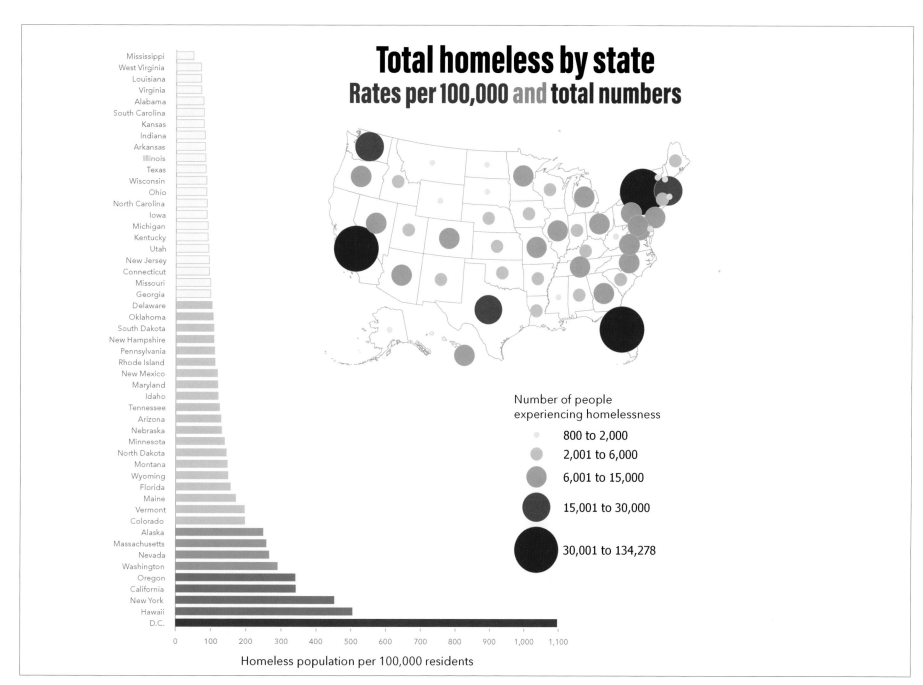

Total homeless by state
Rates per 100,000 and total numbers

Number of people experiencing homelessness

- 800 to 2,000
- 2,001 to 6,000
- 6,001 to 15,000
- 15,001 to 30,000
- 30,001 to 134,278

Homeless population per 100,000 residents

This bar chart shows the number of people experiencing homelessness per 100,000 residents; this rate is highest for Washington, DC, Hawaii, New York, California, and Oregon. Looking at the number of people experiencing homelessness, California has the most, with 134,278, followed by New York with 89,503, Florida with 32,190, and Texas with 23,548. High rates suggest a need for increased prevention. High numbers necessitate investment in resources to support people who are already homeless.

On any given night, more than half a million people experience homelessness in the United States. Tragically, one in five are under the age of 18—that's more than 100,000 homeless children.[1] In addition, the number of homeless youth, defined as people under the age of 25 who are living without a parent or guardian, is increasing.[1,2] Some are runaways; others have been abused or kicked out of their homes. Some are addicted[4] or have aged out of foster care and other juvenile services. As many as 22 percent of homeless girls are pregnant.[5] Most homeless youth live on the street, along river beds, under freeway overpasses, in cars, behind buildings, in parks, or in other makeshift encampments.[6]

Homeless shelters can only accommodate about two-thirds of the total homeless population.[7] They regularly exceed capacity, especially when freezing weather makes living on the street deadly.[8] Homelessness among veterans is also on the rise. Estimates put the number of homeless veterans at more than 40,000, including 3,600 women.[1] As many as 80 percent of homeless veterans struggle with mental illness or addiction.[9] While veterans make up 10 percent of the US adult population,[10] they represent 11 percent of the US adult homeless population.[11]

The human and social tragedy of homelessness costs Americans billions of dollars each year.[13] Orange County, California, for example, estimates that 10 percent of its chronically homeless street population cost the county more than $400,000 per person every year, primarily for jail expenses, ambulance rides, and emergency care.[14] Several jurisdictions have declared states of emergency because of homelessness,[15] including Los Angeles,[16] Anaheim,[17] Seattle,[18] Portland,[19] Tacoma,[20] and the entire state of Hawaii.[21] Solving this problem is essential if we hope to preserve resources, improve the quality of life for everyone, keep families together, and promote the type of community that used to be synonymous with American culture.

Photo by Aaron McLin

"Our city is in the midst of an extraordinary homelessness crisis that needs an extraordinary response. These men, these women, these children are our neighbors."

—Los Angeles Mayor Eric Garcetti[12]

LA's year-round fair weather makes the city attractive for people forced to live on the streets. The city has experienced a surge in homelessness in recent years. (Photo by Clotee Pridgen Allochuku.)

ADDRESSING HOMELESSNESS USING GIS AND SPATIAL ANALYTICS

Complex problems such as homelessness require broad-based, integrative, and data-driven solutions. Spatial analytics help cities, agencies, nonprofits, and other jurisdictions (like sprawling Los Angeles County) synthesize relevant data into a common geographic framework known as GIS. This framework allows stakeholder agencies to address this humanitarian crisis through the coordinated allocation of resources. This chapter focuses on the use of spatial analytics to understand the root causes of homelessness, explore how it is distributed across a region, and target appropriate interventions.

Althoiugh the study area for the workflows in this chapter is the most populated county in the United States, you can apply the same methods to anywhere in the country—or the world—as long as reliable sociodemographic data is available. You can find a companion story map and step-by-step tutorial designed to accompany this chapter at GISforScience.com. These resources provide focused instructions on how to use ArcGIS Pro software to apply the value-based methods described in this chapter.

Using public data for Los Angeles County, the tutorial details these workflows:

* Creating a risk surface for homelessness and examining the spatial patterns of various risk factors
* Mapping the distribution and characteristics of the homeless population
* Weighing options for locating new homeless resources

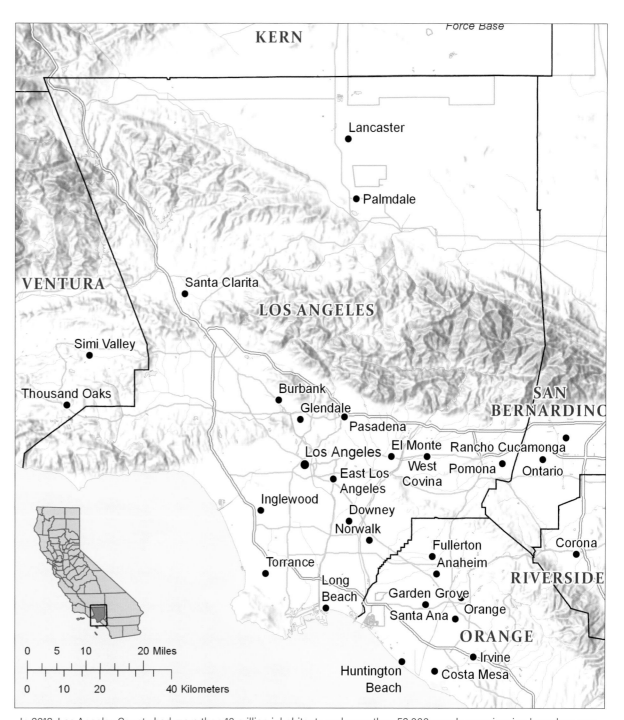

In 2018, Los Angeles County had more than 10 million inhabitants and more than 52,000 people experiencing homelessness.

THE GIS WORKFLOWS

A comprehensive plan to address homelessness will necessarily include a variety of projects, but fundamental to these efforts are two areas of focus: programs aimed at preventing homelessness before it starts and programs designed to aid and support people already experiencing homelessness.

This chapter presents the workflows and the science behind them, supporting both types of programs.

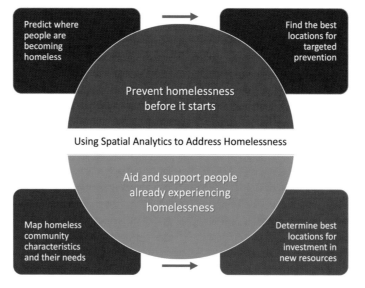

Using Spatial Analytics to Address Homelessness

The Los Angeles County workflow broke the problem into two main areas of focus: (1) preventing homelessness before it starts and (2) providing aid and support for already homeless populations.

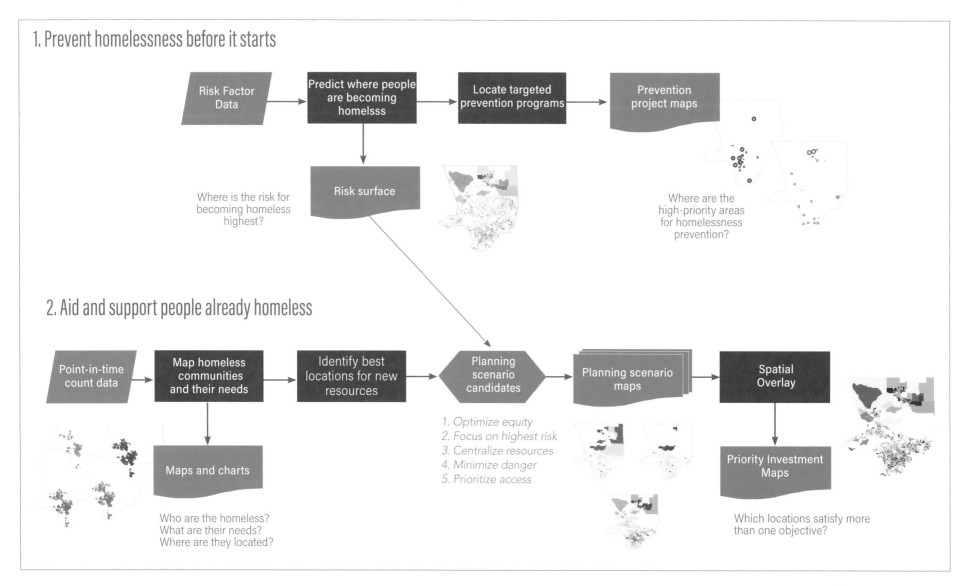

1. Prevent homelessness before it starts

Risk Factor Data → Predict where people are becoming homelsss → Locate targeted prevention programs → Prevention project maps

Where is the risk for becoming homeless highest?

Risk surface

Where are the high-priority areas for homelessness prevention?

2. Aid and support people already homeless

Point-in-time count data → Map homeless communities and their needs → Identify best locations for new resources → Planning scenario candidates → Planning scenario maps → Spatial Overlay

Maps and charts

Who are the homeless?
What are their needs?
Where are they located?

1. Optimize equity
2. Focus on highest risk
3. Centralize resources
4. Minimize danger
5. Prioritize access

Priority Investment Maps

Which locations satisfy more than one objective?

PREDICTING HOMELESSNESS BEFORE IT STARTS

Prevention is much less costly than dealing with the complications of chronic homelessness. Consequently, early intervention is fundamental to any viable plan for combating homelessness. Intervention, however, requires knowing where people are becoming homeless. While counting the number of homeless at a specific point in time[22] identifies the locations of homeless people, these counts don't identify where those people lived when they became homeless. By using spatial analytics and key factors that contribute to homelessness,[23] we can predict where people become homeless.

Predicting the risk of homelessness: A GIS method

Root causes of homelessness arise from a combination of quantifiable risk factors, including poverty, mental illness, substance abuse, domestic violence, and rent disparities. A key to this type of analysis is to gather as much reliable risk factor data as possible for your study area. Consider the end product that you're after: a ranking of all tracts in Los Angeles County according to the presence of homelessness prediction factors, collectively. In the blue box is a list of risk factors for homelessness. You can take four main steps, as shown, to synthesize relevant data into a coherent risk index. More details are provided in a guided tutorial available from this book's companion website at GISforScience.com.

Predicting homelessness: Primary risk factors

Poverty
Number and percentage of households living in poverty. Change in the number of households living in poverty.

Public assistance
Number and percentage of households receiving public assistance income. Change in the number of households receiving public assistance income.

Disabilities
Number and percentage of households with one or more disabled persons.

Access to affordable housing
Number and percentage of households paying more than 50% of their income for rent. Change in the number of households paying more than 50% of their income for rent.

Unemployment
Number and percentage of unemployed population age 16+. Change in the unemployment rate.

Health insurance
Total and percentage of population 18+ with no health insurance.

Veterans and PTSD
Number and percentage of veterans estimated to have PTSD.

Mental illness
Number and percentage of population estimated to suffer from severe mental illness.

Substance abuse
Number and rate (per 100,000 population) of substance abuse incidents.

Domestic violence
Number and rate of domestic violence incidents.

Sources: Esri demographics, National Institute of Mental Health, Los Angeles County GIS Data Portal, U.S. Department of Veteran Affairs National Center for PTSD, Los Angeles County Sheriff's Depatment, and U.S. Census American Community Survey.

STEP 1
Determine the geometry for your analysis (e.g., neighborhoods, census tracts, or zip codes) and gather the data for those features. Notice the tabular attribute data for Los Angeles has been collected at the census tract level and that every tract has risk factor data.

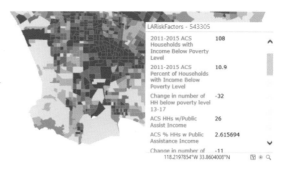

STEP 2
Create a hypothetical census tract and assign it the worst values found across the county for each of the risk factor variables.

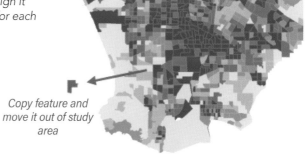

Copy feature and move it out of study area

STEP 3
Run the Similarity Search Tool in ArcGIS Pro. This tool identifies which candidate census tracts are most similar to the hypothetical worst-case tract that we created in step 2 based on a ranking of all the homelessness prediction factors.

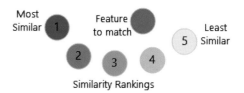

STEP 4
Map the new resulting "similarity layer" (see next page).

PREDICTING HOMELESSNESS RISK

For example, the Los Angeles County risk map shown on this page was created by ranking all census tracts against worst-case risk factor values. For example, one tract (A) has a rate of 12.5 percent for people with serious mental illness (this is the highest rate for that risk factor in the county). In another tract (B), 75 percent of the population is without any kind of health insurance. The largest number of households living below the poverty level for a tract (C) in Los Angeles County is 1,809. The full list of risk factors evaluated includes poverty, unemployment, disabilities, use of public assistance, access to affordable housing, domestic violence, mental illness, post-traumatic stress disorder (PTSD) among veterans, lack of health insurance, and substance abuse. See the online tutorial for detailed steps and data for creating this risk map.

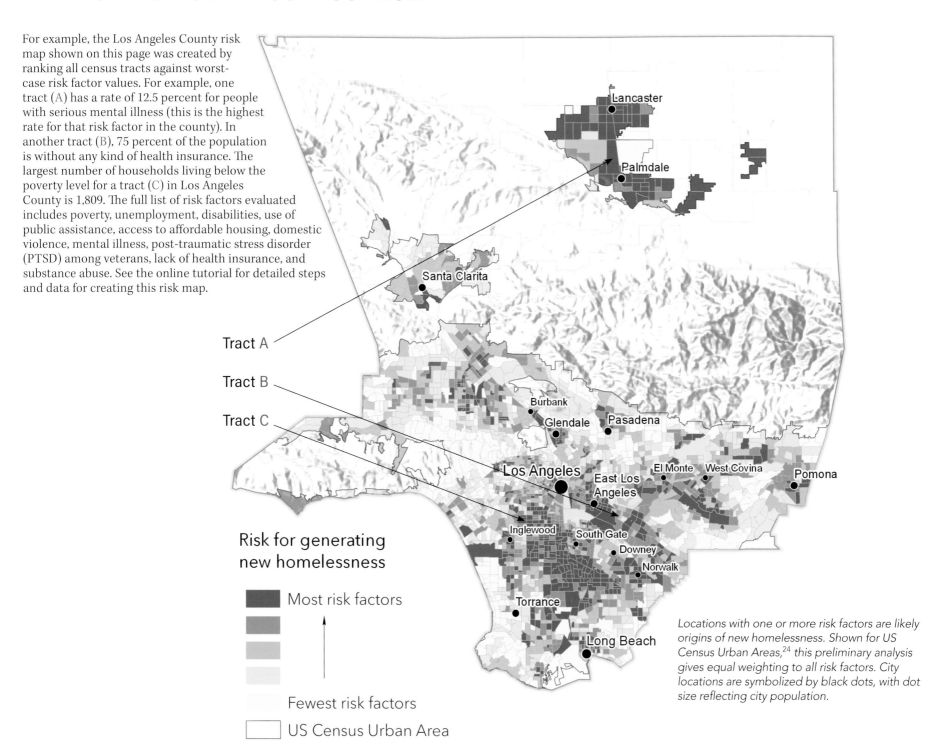

Tract A

Tract B

Tract C

Risk for generating new homelessness

Most risk factors

Fewest risk factors

US Census Urban Area

Locations with one or more risk factors are likely origins of new homelessness. Shown for US Census Urban Areas,[24] this preliminary analysis gives equal weighting to all risk factors. City locations are symbolized by black dots, with dot size reflecting city population.

LOCATING TARGETED PREVENTION PROGRAMS

The risk factors and underlying causes for homelessness are undeniably complex—they are also spatial. The spatial aspect is important because it means spatial analytics can be used to identify where to prioritize targeted programs for preventing homelessness before it starts. One option would extend prevention programs across the entire county (at great cost). A better alternative would use GIS to identify the locations where specific prevention programs are likely to have their biggest impacts.

Veterans, PTSD, and homelessness risk

Preventing veteran homelessness is an important priority in Los Angeles.[25] The next map is symbolized with bivariate renderings to show the relationship between two variables. It uses proportional symbol sizes to show the number of homeless veterans and a dark-to-light color ramp to show the percentage of veterans in each tract estimated to suffer from PTSD.

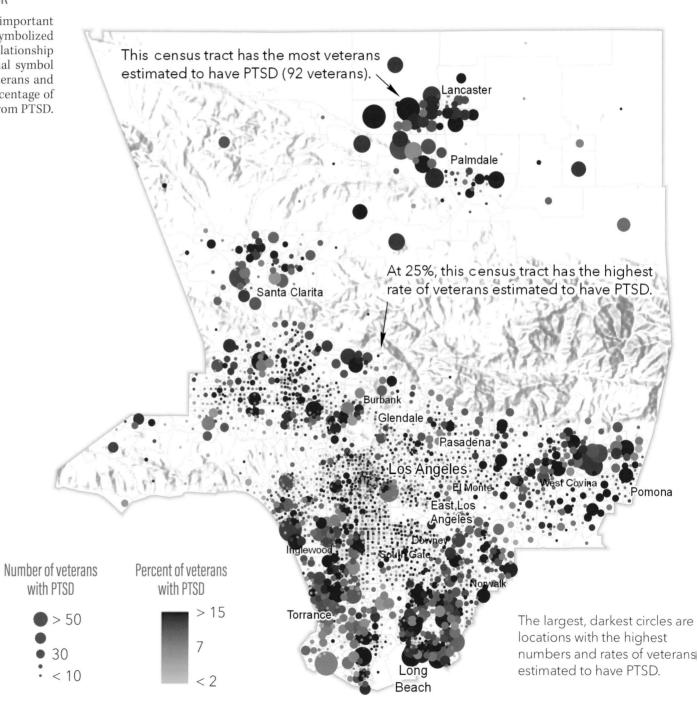

This census tract has the most veterans estimated to have PTSD (92 veterans).

At 25%, this census tract has the highest rate of veterans estimated to have PTSD.

Number of veterans with PTSD

● > 50

● 30

· < 10

Percent of veterans with PTSD

> 15

7

< 2

The largest, darkest circles are locations with the highest numbers and rates of veterans estimated to have PTSD.

The map on this page was created by ranking all tracts against the worst-case values for both the rate and total number of veterans estimated to suffer with PTSD, using the same analysis strategy used to create the risk map on page 89. To compare locations across multiple variables (rates, counts, indices, and so forth), the ArcGIS Pro tool (Similarity Search) standardizes the data and then computes the sum of squared differences between a worst-case target and all other tracts.

The table below shows the top 25 priority locations for programs to prevent veteran homelessness.

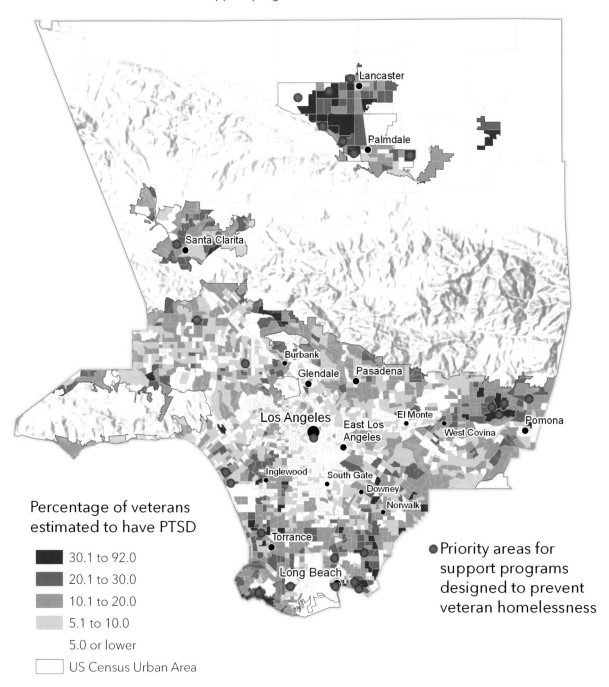

Where should veteran support programs be focused?

Percentage of veterans estimated to have PTSD

- 30.1 to 92.0
- 20.1 to 30.0
- 10.1 to 20.0
- 5.1 to 10.0
- 5.0 or lower
- US Census Urban Area

● Priority areas for support programs designed to prevent veteran homelessness

25 Best Locations fo...Support Centers ×			
TRACT	% Veterans with PTSD	Est Veterans w PTSD	Similarity Rank
901004	10.43084	92	1
901205	9.961686	78	2
702901	9.061489	56	3
572201	10.33058	50	4
401312	9.140768	50	5
910302	8.448276	49	6
576001	10.43689	43	7
111301	11.62791	40	8
295103	10.43257	41	9
670602	6.304079	51	10
920330	8.764941	44	11
910401	9.589041	42	12
577300	10.55409	40	13
620602	10.52632	40	14
401304	9.170306	42	15
123800	11.2462	37	16
910705	11.04478	37	17
900803	9.111618	40	18
401701	10.57143	37	19
910207	9.895834	38	20
573601	9.793814	38	21
577100	9.71867	38	22
206300	6.984127	44	23
400304	8.62069	40	24
276603	11.07595	35	25

Sometimes in GIS, the most important end result is not even a map. In this case, one bit of critical information is the list of targeted tracts shown in this table.

AFFORDABLE HOUSING AND HOMELESSNESS

Access to affordable housing plays an important role in the dynamics of homelessness, not just in Los Angeles, but across the nation,[26, 27] where rising rents and housing policies correlate with rising homelessness.[28, 29] Voucher systems are not entirely effective because they assume there is no lack of housing, only an inability to pay for it.[30] This system doesn't work in Los Angeles, where skyrocketing rents and rock-bottom vacancy rates mean the most effective prevention strategies must include new affordable housing options.[31] The map here shows locations where either a large number or a large proportion of households must devote more than 50 percent of their total income to cover rent. The general rule is that housing should be less than 30 percent of total income[32] to be affordable.

Spatially informed strategies to prevent homelessness are only part of the solution. Additional projects are needed to help existing homeless families and individuals move into permanent housing. These workflows are discussed next.

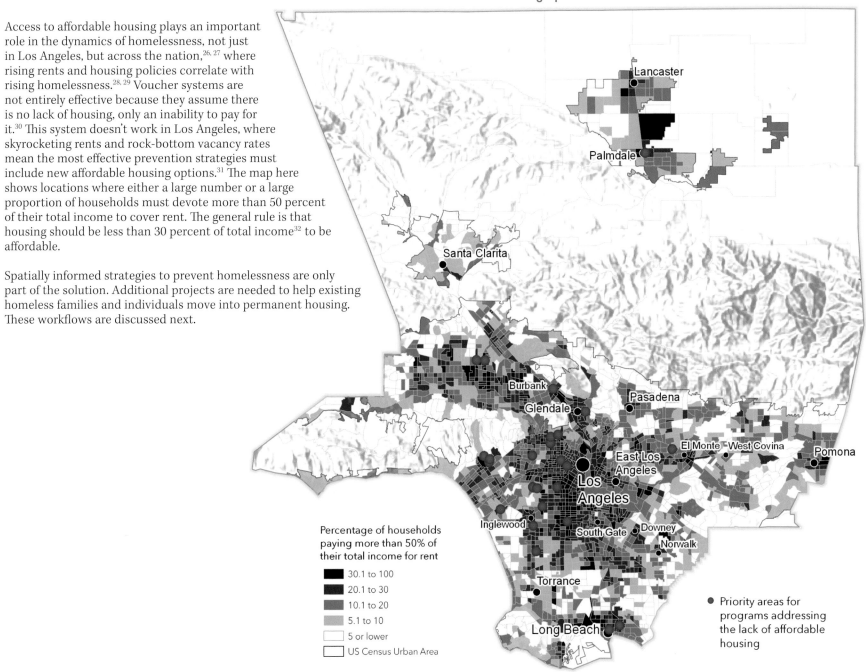

Where are low-cost housing options needed most?

Percentage of households paying more than 50% of their total income for rent

- 30.1 to 100
- 20.1 to 30
- 10.1 to 20
- 5.1 to 10
- 5 or lower
- US Census Urban Area

● Priority areas for programs addressing the lack of affordable housing

HOW HOMELESSNESS DATA IS COLLECTED

Collecting data about people experiencing homelessness

While this chapter focuses on the spatial analytics of homelessness, it is worth noting that geospatial steps happen both before and after analysis. For example, data must be collected prior to any statistical procedures, and interventions will follow from the analytic insights gleaned. Esri provides tools and applications for every step of the process, such as counting the homeless, maintaining an inventory of homeless resources, helping people experiencing homelessness find local services, and engaging the community in reporting homeless activities. For more information about these web applications, see the companion website for this book: GISforScience.com.

The data collection method of gathering information about people already experiencing homelessness is the Point-in-Time Count. This program[33] provides a snapshot of present homelessness counts—a means of collectively understanding and tracking the scope and breadth of homelessness across the nation at the local community level. The US Department of Housing and Urban Development (HUD) requires[34] local jurisdictions receiving federal funding to deliver count results for all sheltered homeless people on an annual basis and for all unsheltered homeless people every two years.

Since 2005 when the counts began, the methodology[35] has primarily involved pens and paper forms. Volunteers venture out on specific nights in January to canvass alleys, parking lots, bridge underpasses, outside stores, and inside shelters. Their goal is to get an accurate census of homeless individuals and learn more about the characteristics of local homeless populations. The government uses this data to set priorities, determine funding levels, and benchmark performance against ongoing efforts to prevent homelessness nationwide.

More recently, pen and paper are giving way to GIS web applications,[36] making the task of delivering an accurate census much easier and producing results more quickly. For example, Spokane, Washington, has already experienced the benefits of going mobile. The city now can conduct its counts biannually, which allows analysis of seasonal and spatial variations in homeless populations.

A volunteer using traditional paper-on-clipboard interviews a homeless person. Equipped with a mobile version of the same survey, his associate collects data in a side-by-side pilot aimed at improving the speed and accuracy of the count. Photo by Zoe Meyers/The Desert Sun.

Applications such as Survey123 for ArcGIS® on tablet devices are changing the nature of point-in-time homeless data collection.

MAPPING HOMELESS COMMUNITY CHARACTERISTICS AND THEIR NEEDS

Once data is collected about people who are experiencing homelessness, important questions can be answered. Who are the homeless? What are their needs? Where are they located? Hot spot analysis applied to the point-in-time count data[37] provides a way to map regions where resources are needed and the form those resources should take. The next maps, for example, show regional concentrations of sheltered and unsheltered homeless people across Los Angeles County.

The Hot Spot Analysis tool in ArcGIS uses the Getis-Ord Gi* statistic. Conceptually, it works by comparing the mean density value of a feature (such as a census tract) and its nearest neighbors to the mean density for the entire study area. If the local mean density value is higher than the global mean density value, and if the difference is statistically significant, the feature is part of a hot spot area (red) on the maps. If the local mean is significantly smaller than the global mean, the census tract is part of a cold spot area (blue).

Knowing where homeless people concentrate is an important first step and should be followed by analyses to understand differences among homeless communities. Why? This analytical work informs better decision making about the type of support and resources needed in each place. The people experiencing homelessness in Venice Beach or Hollywood, for example, have different characteristics (and consequently different needs) than the people experiencing homelessness in Skid Row. This area has the smallest total area but the largest number of homeless people (by far), with 57 percent accessing shelters. In Venice, only 12 percent of the people experiencing homelessness access shelters. Domestic violence is a leading cause of homelessness for women[39] and is the highest risk factor for people experiencing homelessness in Hollywood. The highest risk factor for people experiencing homelessness in Skid Row is serious mental illness. People with untreated mental illness are estimated to comprise one-third of the total homeless population in the United States.[40]

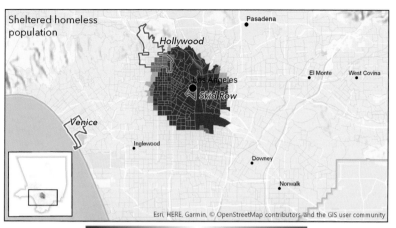

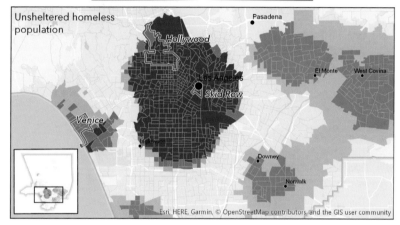

Significant clustering of high densities   Significant clustering of low densities

The largest number of sheltered homeless people are found in and around downtown Los Angeles, especially in the area known as Skid Row.[38] Statistically significant concentrations of unsheltered homeless people are similarly found in Skid Row, with additional clusters in the Venice Beach and Hollywood areas.

Where do the homeless sleep in Los Angeles?

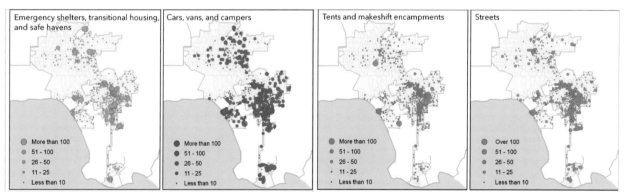

These maps focus on the City of Los Angeles where public data details the living situation of people experiencing homelessness: emergency shelters, transitional housing, or safe havens; cars, vans, or campers; tents and makeshift encampments; or unsheltered on the street.

Charts and other visual analytics allow deeper investigation of homeless community characteristics. A geographic understanding of patterns and relationships is powerful when it can ensure the right intervention for the right people in the right places. Many cities and counties provide clinics, job programs, counseling, shelter opportunities, and food distribution centers to assist individuals and families experiencing homelessness.[41]

Additionally, government agencies are tracking encampments to connect human services with the people needing them and coordinating public works projects to remove trash and ensure healthy environments.[36] Incorporating spatial science into these activities creates a holistic approach to informed action.

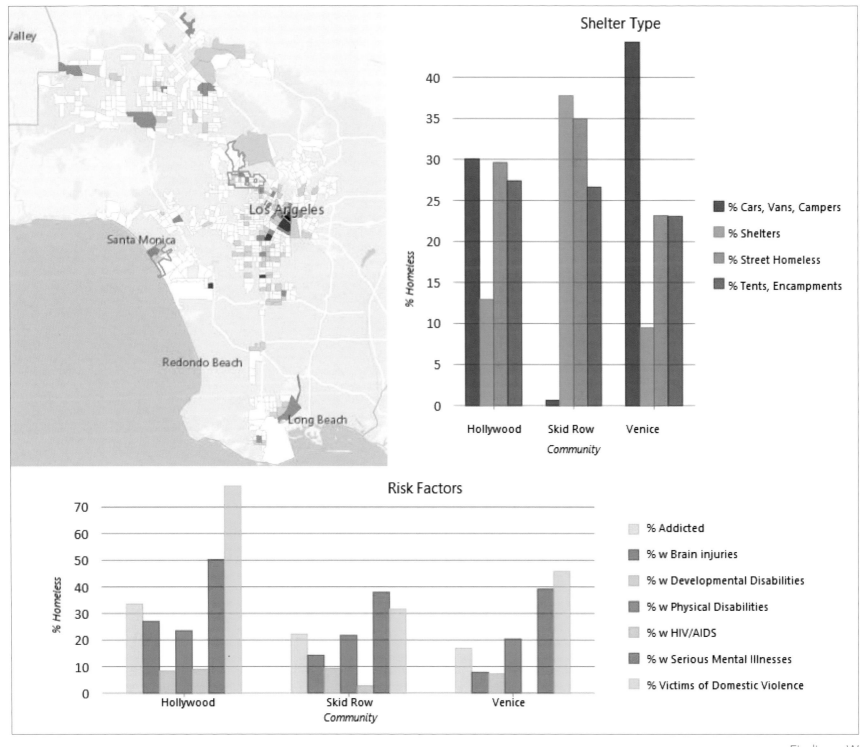

INVESTING WISELY IN NEW RESOURCES

Ask 10 stakeholders and you will likely get 10 different ideas about where to invest in new resources to assist people experiencing homelessness. Potential projects range from expanding rapid rehousing programs, to building new supportive housing units, to implementing targeted homeless encampment interventions.[42] Given the multitude of options, the first step in a data-driven policy approach is to identify a full set of the planning scenarios relevant to stakeholders and local circumstances. Once the different scenarios are defined, they can be mapped.

Mapping different planning scenario objectives provides neutral ground for discussions, encouraging both transparency and collaboration, and increasing focused engagement. The next maps illustrate a variety of possible political and strategic objectives, but they serve only as examples. After each of the planning scenarios are mapped, a final set of maps are created using walk-time distances, and hot spot analysis. These maps identify areas that meet multiple objectives, potentially promoting consensus.

Promote social equity

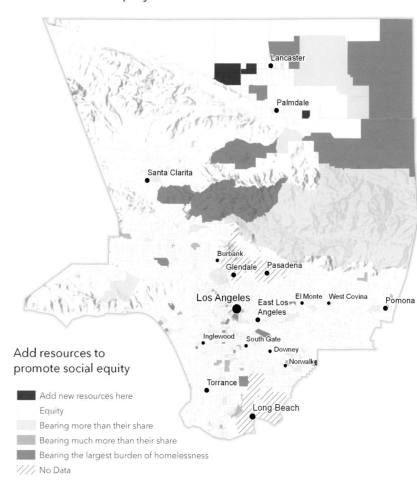

Add resources to promote social equity

- ■ Add new resources here
- □ Equity
- ░ Bearing more than their share
- ▒ Bearing much more than their share
- ▓ Bearing the largest burden of homelessness
- ▨ No Data

Based on the computed social equity index, the red tracts are not bearing their fair share of the homelessness burden. They have a much larger proportion of county residents compared with county homeless.

Modeling social equity

A social equity index may be computed by simply subtracting two ratios: the proportion of all residents in each tract against all tracts and the proportion of all homeless people in each tract to homeless people in all tracts.

 = Social equity index

$p/P - h/H = E$

p is the residential population in the tract
P is the residential population in all tracts
h is the homeless population in the tract
H is the homeless population in all tracts

If a tract has 5 percent of the total population (p/P = 0.05) and 10 percent of the homeless population (h/H = 0.10), E will be a negative value (0.05 – 0.10 = -0.05), indicating a tract that is bearing more than its share of the burden of homelessness. Similarly, in tracts with a larger proportion of residents than people experiencing homelessness, E will be positive, indicating a tract that is bearing less than its share of the burden.

Prioritize assistance for the newly homeless

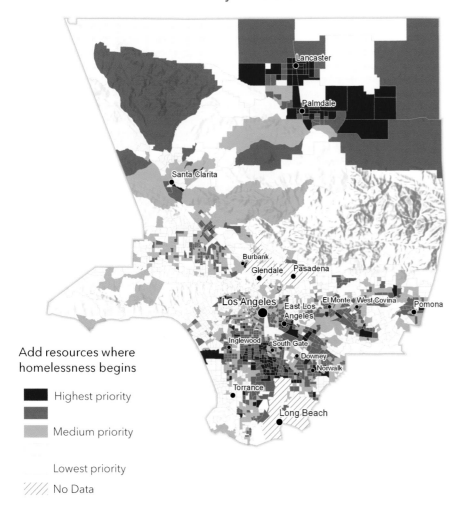

Add resources where homelessness begins

- ■ Highest priority
- ■ Medium priority
- □ Lowest priority
- ▨ No Data

If resources are available where people become homeless, they are more likely to remain in or near their own communities where their children attend school and where they are likely to have a broader set of personal and social resources. The highest priority risk areas where people are likely becoming homeless were obtained from the map of risk rankings. Providing resources in high-risk areas is the strategy being promoted by the mayor of New York.[43]

Centralize homeless resources

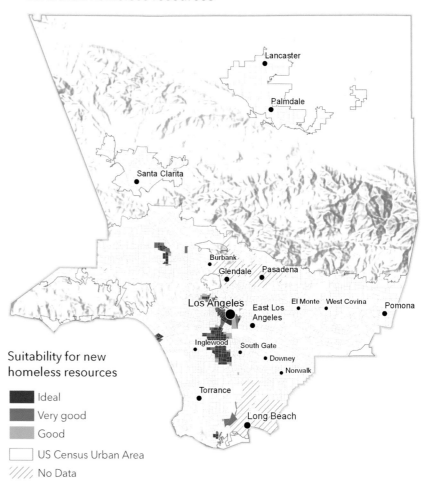

Suitability for new homeless resources

- ■ Ideal
- ■ Very good
- □ Good
- □ US Census Urban Area
- ▨ No Data

Consolidating new resources into geographically centralized locations is the strategy being adopted by the City of San Francisco.[44] This planning scenario promotes adding new resources to locations that currently have resources.

Focus on the most dangerous and costly locations

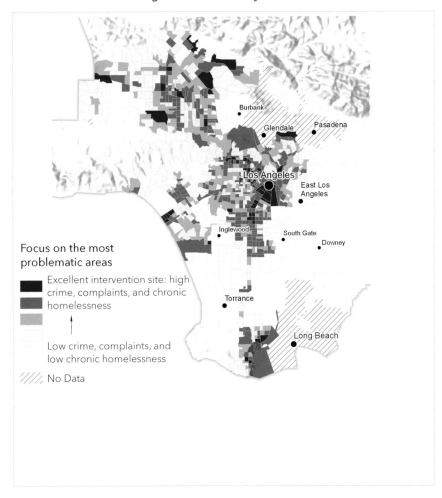

Focus on the most problematic areas

■ Excellent intervention site: high crime, complaints, and chronic homelessness

Low crime, complaints, and low chronic homelessness

//// No Data

By evaluating high numbers of 311 calls,* crime incidents perpetrated by or against homeless individuals, and high numbers of chronically homeless people, the most vulnerable homeless populations can be identified. People in those locations become candidates for rapid-response, focused interventions aimed at getting every homeless person the resources they need to move out of homelessness permanently. Research indicates a small portion of the homeless population, often people who are chronically homeless, use most of the money targeted for homelessness. Addressing the locations with the most vulnerable and most costly homeless people first will have the biggest impact on reducing costs.[45]

*311 is a phone hotline number that citizens can use to get local resource information or to report a problem in their community.

Improve accessibility

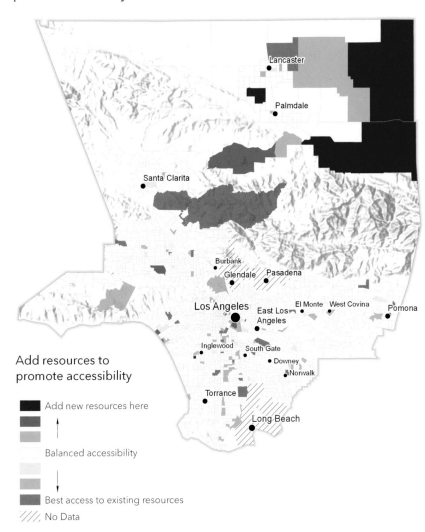

Add resources to promote accessibility

■ Add new resources here

Balanced accessibility

Best access to existing resources

//// No Data

Another possible planning scenario focuses on improving accessibility by putting new facilities in tracts where existing homeless people live and where few resources are available. Identifying these areas considers supply and demand.

Spatial data has countless uses for specific policy initiatives. A Housing First solution, for example, might prioritize locations for new permanent housing or for converting existing housing into homes for people experiencing homelessness.

Once all planning scenarios have been mapped, overlaying them takes decision making to the next level. A composite map can reveal whether locations meet multiple objectives and, therefore, can offer the highest impacts.

Spatial overlay analysis is performed on census tracts, and while this approach is convenient and effective, homelessness, like many other social problems, does not honor administrative boundaries.

To address this and improve the final map, we can use hot spot analysis. This kind of analysis identifies spatial clustering of the number of objectives met in each tract within a 20-minute walking distance. The result is larger areas, giving decision makers more flexibility for resource development.

County officials can use a map like this one to select candidate areas for locating new homeless resources such as shelters, food distribution centers, and a variety of legal, job training, medical treatment, and counseling facilities.

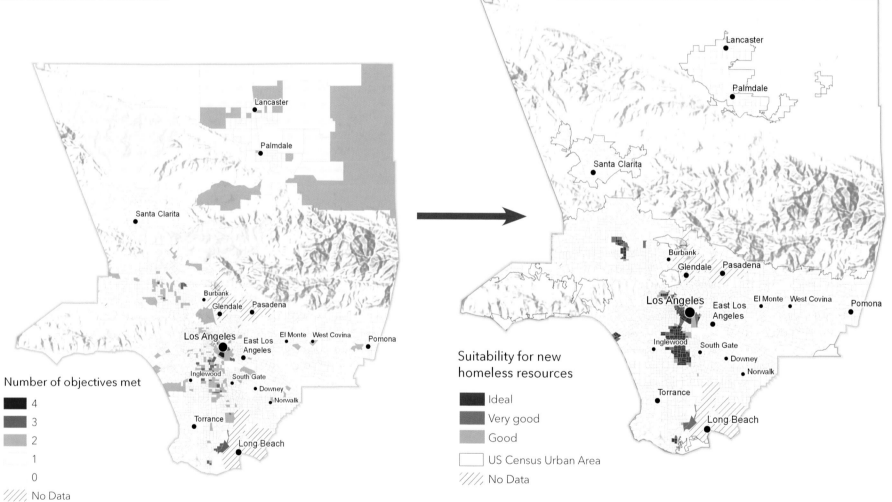

Number of objectives met

- 4
- 3
- 2
- 1
- 0
- //// No Data

Suitability for new homeless resources

- Ideal
- Very good
- Good
- [ ] US Census Urban Area
- //// No Data

Spatial overlay analysis synthesizes political and strategic objectives into candidate regions where investments are likely to have the biggest impact on the lives of people experiencing homelessness.

ENSURING SUCCESS

Critical to any action plan is the careful assessment of program success. Knowing what works and what doesn't helps facilitate effective investment of limited resources. In the graphic below, a variety of statistics about the people experiencing homelessness are tracked.

A coordinated effort across agencies and departments will be needed to register all homeless services and to track both utilization and capacity. With linked records, every encounter with a homeless person becomes an opportunity to quantify the effectiveness of services rendered. Metrics to track program impacts should answer questions such as these:

1. Is the average number of days a person remains homeless going down?
2. Is the percentage of chronically homeless people decreasing?
3. Are outreach programs (e.g., substance abuse, domestic violence, unemployment, or institutional discharge) assisting a larger number of people? Are fewer people experiencing homelessness as a result of these programs?
4. Has every veteran experiencing homelessness been placed in permanent housing?
5. Does emergency shelter and housing capacity match the number of newly homeless each month?

More difficult, but also more important, is the fundamental task of tracking homelessness prevention. Preventing homelessness requires understanding which communities are at risk and then identifying key risk factors so that targeted interventions can be administered. Ultimately, the goal is to be able to quantify the cost of prevention, the cost of moving a homeless person from the street into permanent housing, and the cost of ongoing support for people already experiencing homelessness. These details are needed to optimize return on investment.

Homelessness intersects many aspects of our society, including policing, public works, infrastructure, public health, education, employment, and the economy. Without doubt, cities, agencies, and GIS analysts will need to leverage all tools available to effectively address the homelessness crisis in America.

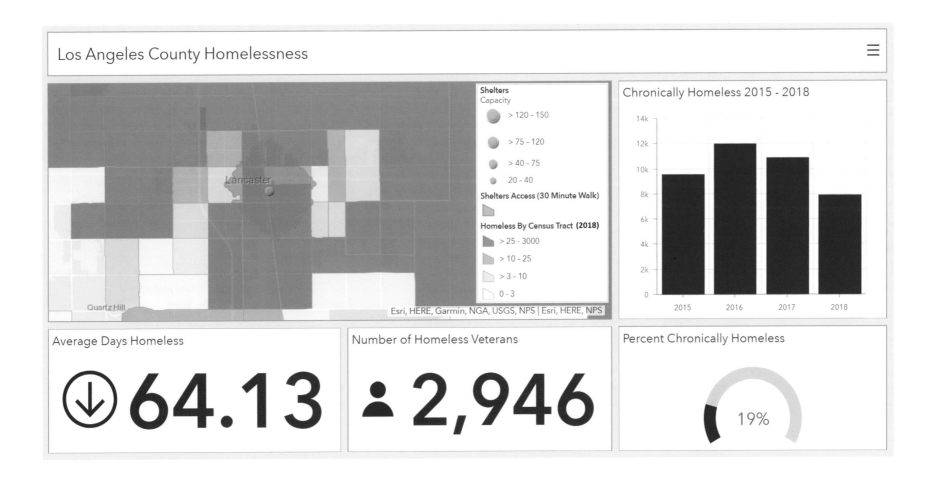

ENDNOTES

1. The US Department of Housing and Urban Development, Office of Community Planning and Development, *The 2017 Annual Homeless Assessment Report (AHAR) to Congress* (December 2017).

2. Teresa Wiltz, *A Hidden Population: Youth Homelessness Is on the Rise* (July 7, 2017): The PEW Charitable Trusts. Research and Analysis, Stateline.

3. M. H. Morton, A. Dworsky, J. Matjasko, S. R. Curry, D. Schlueter, R. Chavez, and A.F. Farrell, "Prevalence and Correlates of Youth Homelessness in the United States," *Journal of Adolescent Health* (2018): 14–21.

4. D. M. Santa Maria, S. C. Narendorf, and M. B. Cross, "Prevalence and Correlates of Substance Use in Homeless Youth and Young Adults, *Journal of Addictions Nursing* 29, no. 1 (2018): 23–31.

5. National Conference of State Legislatures, *Homeless and Runaway Youth.*

6. Hugo Aguas, "Homeless Demography in Los Angeles County," *Environmental Criminology: Spatial Analysis and Regional Issues* 20 (2018): 69–86, Emerald Publishing Limited.

7. Samantha Raphelson, "Shelters Reach Capacity in Cold Weather As Homeless Population Rises," NPR, *Here and Now Compass, Special Series* (January 9, 2018).

8. Grace Guarnieri, "Freezing Weather Threatens Lives of Homeless Americans as Shelters Nationwide Are Pushed Past the Limit, *Newsweek* (January 3, 2018): US Edition.

9. D. A. Smelson, M. Chinman, G. Hannah, T. Byrne, and S. McCarthy, "An Evidenced-Based Co-occurring Disorder Intervention in VA Homeless Programs: Outcomes From a Hybrid III Trial," *BMC Health Services Research* (2018).

10. Kristen Bialik, "The Changing Face of America's Veteran Population," *Factank* (November 10, 2017): Pew Research Center.

11. National Coalition for Homeless Veterans. Background and Statistics.

12. Doug Smith and Gale Holland, "L.A. County Homelessness Jumps a 'Staggering' 23% as Need Far Outpaces Housing, New Count Shows," *Los Angeles Times* (May 31, 2017).

13. United States Interagency Council on Homelessness, *Investing in the End of Homelessness: The President's 2017 Budget* (2017).

14. David Snow and Rachel E. Goldberg, with The United Way, Orange County, Jamboree, and University of California, Irvine, *Homelessness in Orange County: The Costs to Our Community* (June 2017).

15. J. G. Wogan, "Why Governments Declare a Homeless State of Emergency," *Governing the States and Localities, Health & Human Services* (November 10, 2015): accessed 5/25/2018.

16. Dana Ford, "Los Angeles Declares 'State of Emergency' on Homelessness, *CNN* (September 23, 2015): accessed 5/25/2018.

17. J. Pimentel and T. Walker, "After Anaheim Declares State of Emergency, Santa Ana River Homeless Wonder What's Next," *Orange County Register* (September 28, 2017): accessed 6/13/2018.

18. Lydia O'Connor, "Seattle Declares State of Emergency on Homeless Crisis, *HUFFPOST*, Impact (January 9, 2017).

19. Molly Harbarger, "A New Vision for Portland's Homeless Crisis a Year after 'State of Emergency,'" *OREGONLIVE/The Oregonian* (February 20, 2017): Portland News, accessed 5/25/2018.

20. King5 News staff, "Tacoma Declares Homeless State of Emergency," *King5 News* (May 9, 2017).

21. Justin Worland, "Hawaii Governor Declares State of Emergency Over Homeless Plight," *Time Magazine* (October 18, 2015).

22. LAHSA, *Greater Los Angeles Homeless Count (2017).*

23. National Coalition for the Homeless, *Homelessness in America. Building a Movement to End Homelessness.*

24. The US Census Urban Areas Showing Residential, Commercial, and Other Non-Residential Urban Land Uses. The Census Bureau Recognizes Urban Areas with at Least 50,000 People and Urban Clusters with 2,500 to 50,000 People.

25. Dakota Smith and Doug Smith, "Garcetti Says LA Can End Street Homelessness in a Decade," *Los Angeles Times* (March 21, 2018).

26. National Alliance to End Homelessness, *Unsheltered Homelessness; Trends, Causes, and Strategies to Address* (June 2017).

27. Tanvi Misra, "Every US County Has an Affordable Housing Crisis," *CITYLAB* (April 27, 2017).

28. Charley Willison, "Shelter from the Storm: Roles, Responsibilities, and Challenges in United States Housing Policy Governance," *Health Policy* 121 no. 11 (November 2017): 1113-123.

29. Devan Kaney, "Rising Rent in America's Largest Cities is Increasing Homelessness," *NBC26* (June 12, 2018).

30. Jon Erickson and Charles Wilhelm, *Housing the Homeless* (New York: Routledge, 2011).

31. Elijah Chiland, "7 Ways LA Tried to Solve Its Housing Crisis in 2017: Will Any of This Bring Down the Cost of Rent?" *Curbed Los Angeles* (December 20, 2017).

32. Kathleen Elkins, "Here's How Much of Your Income You Should Be Spending on Housing," *CNBC Make It* (June 6, 2018).

33. Corporation for National and Community Service, *National Point in TIme Count of People Experiencing Homelessness.*

34. HUD Exchange, *PIT and HIC Guides, Tools, and Webinars* (2018).

35. Sarah Holder and Linda Poon, "The Tech That's Changing How Cities Help the Homeless," *CITYLAB* (May 31, 2018).

36. Zack Quaintance, "Aurora, Colo., Deploys New Tech to Get More Accurate Homeless Population Count," *Government Technology* (June 9, 2017): Health and Human Services.

37. The point-in-time data used for analysis was collected in January 2017 for the Los Angeles Continuum of Care (CoC). The CoC region encompasses most of the county, but excludes Pasadena, Glendale, and Long Beach.

38. Hadley Meares, "The Early Days of Skid Row, Once Known as 'Hobo Corner,'" *Curbed Los Angeles* (December 14, 2017).

39. National Law Center on Homelessness & Poverty (NLCHP), *Housing and Homelessness in the United States of America* (2014).

40. Treatment Advocacy Center, *Serious Mental Illness and Homelessness* (September 2016): accessed 5/29/2018.

41. Hannah Moss, *Responding to Homelessness in Crisis Mode.* Govloop (January 19, 2018).

42. Los Angeles County, "Here's How LA County Could Spend $355 Million a Year to Fight Homelessness." *Homeless Initiative (*2017).

43. City of New York Press Office. 2017, February 28. De Blasio Administration Announces Plan to Turn the Tide on Homelessness with Borough-Based Approach: Plan Will Reduce Shelter Facilities by Forty Five Percent. NYC, *The Official Website of the City of New York, Office of the Mayor.*

44. Department of Homelessness and Supportive Housing, *San Francisco Navigation Centers: A Housing Focused, Welcoming, Short-Term Shelter Model,* January 2017.

45. Halil Toros and Daniel Faming, Prioritizing Which Homeless People get Housing Using Predictive Algorithms, *Economic Roundtable* (April 29, 2017).

Aileen Buckley designed most of the maps in this chapter. Tanya Bigos created the dashboard on page 102. Jim Herries produced the map on pages 92 and 93. Chris Vaillancourt provided valuable cartographic assistance.

RESTORING COASTAL MARINE HABITATS

Mapping the bond between people and nature, scientists are using geospatial technologies to build coastal resilience by addressing rising sea levels and other impacts of climate change.

By Zach Ferdaña, Laura Flessner, Matt Silveira, and Morgan Chow, The Nature Conservancy; Tom Brouwer, FloodTags, and Omar Abou-Samra, American Red Cross

Nature reasserting its presence at a coral reef breakwater structure in Grenville Bay, Grenada. By understanding nature, marine scientists and engineers are devising ways to reverse the effects of habitat degradation and coastal erosion. Photo credit: Tim Calver

USING NATURE-BASED SOLUTIONS

Coastal communities around the world increasingly see and feel the impacts of climate change. More intense and frequent storms and hurricanes, coupled with rising seas, are changing the land and seascape and dramatically forcing cities, organizations, and nations to reconsider how and where to invest their coastal resources. Storms and floods affect hundreds of millions of people, important infrastructure, and tourism, causing significant losses to local and national economies and livelihoods. For more than 60 years, The Nature Conservancy (TNC), has been well known for acquiring, conserving and restoring coastal habitats and ecosystems around the world to protect nature now and for future generations.

Today, the Conservancy increasingly focuses on enhancing environmental protection to address disaster risk reduction (DRR). The United Nations defines this as any strategy which aims to reduce the damage caused by natural hazards like earthquakes, floods, droughts and cyclones, through an ethic of prevention. Since 2007, TNC has focused its attention on nature-based solutions, the conservation and restoration of degraded oyster and coral reefs, tidal marshes, and coastal mangroves as viable DRR alternatives. Working in public-private partnerships, the group has coined the term *Coastal Resilience*, the combined efforts to help address the devastating effects of climate change and natural disasters.

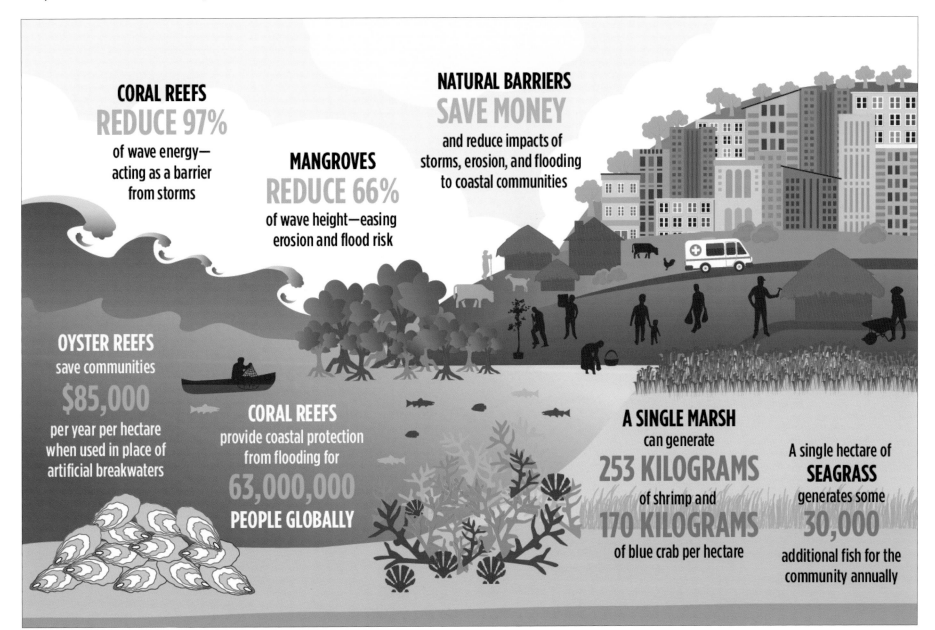

CORAL REEFS
REDUCE 97%
of wave energy—
acting as a barrier
from storms

MANGROVES
REDUCE 66%
of wave height—easing
erosion and flood risk

NATURAL BARRIERS
SAVE MONEY
and reduce impacts of
storms, erosion, and flooding
to coastal communities

OYSTER REEFS
save communities
$85,000
per year per hectare
when used in place of
artificial breakwaters

CORAL REEFS
provide coastal protection
from flooding for
63,000,000
PEOPLE GLOBALLY

A SINGLE MARSH
can generate
253 KILOGRAMS
of shrimp and
170 KILOGRAMS
of blue crab per hectare

A single hectare of
SEAGRASS
generates some
30,000
additional fish for the
community annually

Some of the most effective strategies for reducing flood risk are rooted in restoring natural infrastructure that has always protected coastal environments.

WHAT IS COASTAL RESILIENCE?

Resilience is the capacity of a system to absorb disturbance and still retain its basic function and structure. With natural disasters and climate change, this definition should be expanded to recognize that our social and ecological systems must be nimble or adaptable in how we manage and sustain our communities and the natural world. This area is where GIS science and geospatial design, or geodesign, can help support this expanded concept of resilience. Geodesign is a design and planning method that tightly couples the creation of design proposals with impact simulations informed by geographic contexts (Flaxman 2010). Simply put, geodesign informs resilience to support communities in adaptive planning. The goal aims to empower stakeholders in this Coastal Resilience adaptive-management approach to help communities become more climate resilient using natural ecosystems as part of the solution.

The Coastal Resilience program led by TNC includes projects undertaken in the past decade in the Caribbean, Southeast Asia, Australia, Mexico, Central America, and in most US coastal states to continually examine nature's role in the reduction of coastal flood risk. The general ideas of an ecosystem-based approach are considered globally, but the use of GIS and online mapping decision-support systems allow pinpointed conservation and climate-resilience planning.

Geodesign supports this expanded concept of resilience. The Coastal Resilience approach allows stakeholders take a series of steps to reduce the ecological and socioeconomic risks of coastal hazards:

1. Assess risk and vulnerability to coastal hazards by including current and future storms and sea level rise.
2. Identify potential solutions for reducing risk that benefit social, economic, and ecological systems.
3. Act to help communities develop and implement these solutions.
4. Measure solution effectiveness to ensure that efforts to reduce risk through restoration and adaptation are successful.

The backbone of the approach is communicated through a decision-support system includes an online mapping tool that helps planners, government officials, and communities develop risk reduction, restoration, and resilience strategies. A data-viewing platform and suite of web-responsive apps are designed and tailored to meet specific planning needs, including coastal management policies, post-storm disaster decision making, community assessments, hazard mitigation plans, and cost-effectiveness evaluations.

The capacity of natural ecosystems to provide coastal protection and other services, including food production, water purification, carbon sequestration, and tourism and recreation, is critical yet highly variable (Hale et al. 2009; Hale et al. 2011; Spalding et al. 2012; Arkema et al. 2013). The science of nature- and ecosystem-based solutions in reducing coastal community flood risk is growing rapidly. Designers of coastal resilience strategies refer collectively to these nature-inspired strategies as "green infrastructure."

In contrast, built or "gray" (e.g., concrete) infrastructure such as seawalls and dikes helps safeguard coastal communities from flooding events but is often cost-prohibitive, massive building efforts that can take decades to get funded and built. Gray infrastructure also has socioeconomic and ecological drawbacks. These projects can block public access to natural areas, cause undesired erosion of beaches and coastal habitats, and interrupt natural coastal processes that prevent habitat migration and natural adaptation (Flessner et. al 2016). Natural adaptation can include green infrastructure, such as the creation of new beaches and dunes, marshes and mangroves, and coral and oyster reefs—so-called "soft-engineered" solutions, also called *nature-based* solutions.

Coupled with an increased effort to identify where coastal ecosystems can reduce flood risk, there has been a growing awareness of the need to develop collaborative partnerships and programs between humanitarian and environmental organizations. TNC, through its growing body of work in climate adaptation and mitigation, has formed a partnership with the International Federation of the Red Cross (IFRC). Explicitly incorporating the conservation of ecosystems into the risk reduction equation gives communities the information they need to find the most cost-effective and multi-beneficial adaptation solutions (Ferdaña et al. 2010; Ferrario et al. 2014; Groves et al. 2012; Halpern et al. 2012; Narayan et al. 2016; Whelchel and Beck 2016). Through collaboration on specific projects in the Caribbean and Southeast Asia that will be highlighted in this chapter, this initial round of work aims to help both TNC and Red Cross identify the most effective ways to integrate ecological disaster risk-reduction efforts within the humanitarian program cycle.

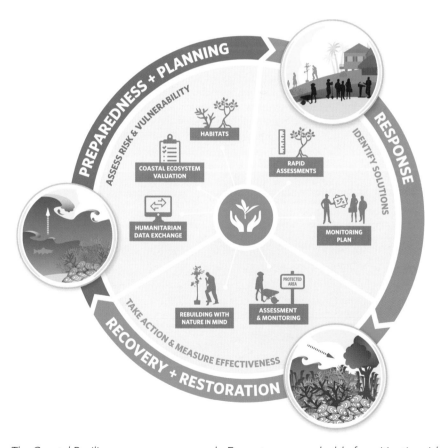

The Coastal Resilience program approach: Ecosystems are valuable for mitigating risk and supporting development and humanitarian goals. Incorporating conservation principles in risk-reduction plans gives communities the information they need to lower costs, discover benefits, and find creative solutions.

CASE STUDY: AT THE WATER'S EDGE

For decades on Grenada's eastern shore, climate change has adversely affected the communities of the Grenville Bay area (Telescope, Grenville, Soubise, and Marquis). Storms, rising seas, and changing temperatures threaten lives and property. A drastic decline in the health of the fringing coral reef and the overharvesting of mangroves exacerbate issues such as flooding and severe coastal erosion.

In 2012, TNC partnered with community members, nongovernmental organizations (NGOs), and government agencies to address these and other environmental challenges through the implementation of strategies aimed at building artificial coral reefs. (The word *artificial* is misleading: over time, what starts as artificial structures eventually become part of nature.) Using resilience and geodesign principles, this effort combined scientific data modeling and interactive mapping tools with traditional community knowledge. The goal was to answer this question: Can complex modeling of waves and wind, coupled with an online mapping tool, effectively support local communities and government ministers?

Accessing tools to identify solutions

Our initial efforts began with the development of the web-based, Coastal Resilience decision-support tool to help government planners and managers further understand specifically where they were most vulnerable. The tool enabled users to test scenarios using census data, data on storm surge occurrences, rise in sea level, existing natural resources, and other community assets, while developing practical restoration solutions that benefit people, nature, and infrastructure. The resulting maps and models helped users layer socioeconomic and vulnerability data to identify and compare areas of concern with areas of potential resilience. Despite national-level advancements in natural sciences and spatial analyses, high-tech, computer-based tools were inaccessible to many local decision makers, who are an integral part of the resilience approach. To further engage with the impacted communities, TNC teamed with local organizations, including the Grenada Fund

for Conservation and Grenada Red Cross Society. Participatory 3D physical maps are a widely accepted method for capturing local knowledge and values through a 3D model of a specific site. By creating a large model of Grenville Bay, nearly 500 residents visualized their communities and highlighted important sites of cultural, economic, and historic value. The map captured and incorporated their knowledge into a dataset that informed the vulnerability assessment process. The dataset also informed the process of selecting cost-effective and culturally appropriate ecosystem-based adaptation strategies that would benefit the community and coastal landscape. Spatial indices such as adaptive capacity, social sensitivity, and critical infrastructure were created, identifying Grenville Bay as the target site because of its high vulnerability to coastal hazards and low adaptive capacity.

To identify and implement potential nature-based solutions in Grenville Bay, TNC collaborated with IH Cantabria to generate 3D digital models from more than 60 years of wave data. The models incorporated different coral reef breakwater design scenarios to understand the best position and placement of breakwaters for reducing coastal erosion and flood risk. The scenarios helped government officials and community leaders identify where breakwaters enhance shoreline protection and thereby improve social, economic, and cultural systems within the Grenville Bay communities. By coupling model results fed into the online decision-support tool with community participation, we empowered stakeholders to select options that best align with their community vision (i.e., prioritizing beach access and enhancing fisheries).

"We are able to bring people together with nature, making them more resilient and able to adapt and combat the issue of climate change and even disaster hazards that would affect the community at the water's edge."

—Terry Charles, Grenada Red Cross Society

Grenville Bay, Grenada, location of the coral reef breakwater project.

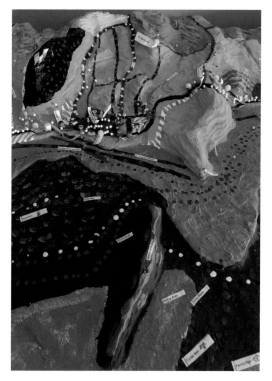

3D cardboard model developed and labelled by stakeholders and government officials highlighting Grenada's land and water resources.

Taking action

The community chose a solution that included the pilot installation of low-crested, hybrid breakwater structures along a degraded reef flat. To maximize economic and ecological benefits, the structures were engineered to meet these design criteria:

- Withstand hurricane wave forces
- Have a minimum lifespan of 30 years
- Promote coralline algae and coral growth
- Provide habitat for fish, lobster, and other commercially important species
- Use local material in their construction
- Be installed by local workers

In 2015, TNC worked with the communities to construct and deploy an initial 30 meters of constructed rebar baskets filled with rocks and hollow concrete blocks, using 15 local fishermen and a local commercial dive operator to transport the baskets. Coral fragments, termed *fragments of opportunity*, were affixed to the installed baskets.

Measuring effectiveness

The project team has monitored the structures to assess their ability to withstand forces, such as wave energy and sand blasting, and their capacity to host crustose coralline algae and other species. Reports indicate that most of the structures have crusted over with coralline algae, which adds to their strength and security. Coral recruits are prevalent, with fish and lobster preferring the steel rebar cages filled with cavernous blocks, as compared with the structures made with only rocks. This pilot structure is now helping inform the build-out of an additional 20 structures along the reef flat, totaling 300 meters.

The At the Water's Edge (AWE) project demonstrates that governments and communities of small island states can enhance their resilience to climate change by protecting, restoring, and managing their marine and coastal ecosystems through capacity building and adaptation. This flagship project between TNC and the Red Cross strengthened our understanding of community resilience using coral reef engineering solutions, setting the stage for more collaboration globally.

Coral reef growth on the breakwater structure one year after implementation, in 2015. The structures are designed to restore the wave-breaking function of the reef, and therefore reduce wave energy reaching the shore, which typically exacerbates coastal erosion and flooding. (Photo by Tim Calver.)

Divers survey the pilot breakwater structure for coral reef growth and structural integrity. (Photo by Tim Calver.)

CASE STUDY: RESILIENT COASTAL CITIES

The Resilient Coastal Cities project builds local coalitions to identify and enhance community resilience. Led by the American Red Cross and the Global Disaster Preparedness Center (GDPC), the project builds on existing approaches to community assessment, problem-solving, and outreach on preparedness measures developed by the local Red Cross network in Indonesia. The Red Cross increasingly integrates environmental management within the humanitarian program cycle (preparedness, response, and recovery) in partnership with TNC, building green approaches to disaster-risk reduction.

In this use case, TNC's Coastal Resilience program leveraged geospatial technology to assess risk and vulnerability from flooding, identify nature-based adaptation solutions, take conservation and restoration action, and measure effectiveness of solutions. Here TNC studied the role of mangroves in protecting the coast from flooding in Semarang, Indonesia.

Social media tools for real-time flood mapping

The coalition identified FloodTags, a social enterprise based in the Netherlands, as a critical component of the framework for identifying when and where floods are happening or have occurred. By monitoring flood-related information from social media and consuming flood reports through direct messaging apps, they detect floods in real time and provide practical information about floods. For this project, FloodTags analyzed flood reports from Twitter, which is popular in Indonesia, to detect flood events for Semarang and other places along the north coast of Central Java.

Examples of flood-related tweets for the north coast of Central Java, Indonesia.

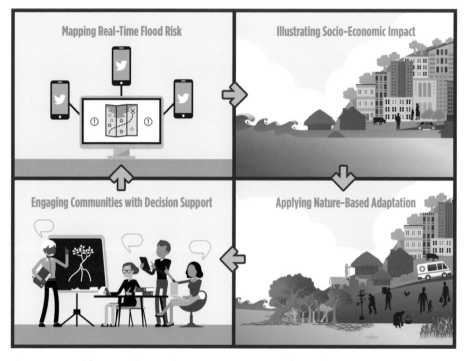

The geospatial framework aimed to combine real-time flood risk with community vulnerability and natural resource mapping into a replicable and scalable approach to help identify possible adaptation solutions for city planning.

"We need to stay relevant and progressive, thinking creatively rather than just doing things how we have always done them."

—Omar Abou-Samra, director, Global Disaster Preparedness Center (GDPC)

Workflow

This analysis included the following steps:

1. Collecting flood-related tweets from the Twitter streaming API, using Indonesian (Bahasa) flood-related keywords.
2. Tweets were classified using a machine-learning algorithm, extracting water depths from them, detecting locations from the text, and determining whether extreme rainfall occurred at these locations.
3. Flood events were detected by comparing the volume of tweets for each location to historical statistics.
4. Enriched flood reports and events were distributed through an API in real time.

The coalition derived the locations from the text of the tweets. Since the location of the flood may not be the same as the location of the person tweeting about it, the GPS coordinates of a tweet were not directly used. Instead, the location names mentioned in the body text were geo-parsed using the Geonames database (see: Geonames, n.d.). Additional metadata from the tweets, including the location of the user, their time zone, GPS coordinates of the tweet, and the names of other locations referenced in the text were used to find the most likely location (in case a location name occurs in multiple places) (see De Bruijn et al. 2018). For this project, locations were derived up to level 4 administrative areas (called Kelurahans or Desas, small neighborhoods) on the north coast of Central Java.

Satellite-derived rainfall data (JAXA 2018) was also used to determine whether extreme rainfall occurred at the derived locations. For this analysis, a history of GSMAP data, dating back to 2008, was analyzed to derive rainfall statistics for each raster cell. Based on these statistics, the percentile of rainfall amounts was determined in near real time, which is an indicator of extreme rainfall. Each tweet with a derived location was then assigned the highest hourly rainfall percentile that occurred over the 24-hour period prior to the tweet being posted.

The tweet classification algorithm was trained using a system called the *Relevancer* (**Hürriyetoğlu** et al. 2016). TNC staff in Indonesia used this system to divide batches of similar tweets into "flood," "flood-related," "mixed," and "irrelevant" classes. After training the algorithm, it has since been implemented to classify tweets in real time.

To detect events, the number of incoming tweets for level 2 administrative areas (including underlying level 3 and 4 areas) were compared with historical statistics for these areas. In case a spike in the number of tweets is detected for an area, the system will flag this increase as a flood event.

The individual flood reports and the events are distributed through the FloodTags API. As floods occur, flood-event data from tweets will automatically update the Resilient Coastal Cities web app, which is connected to the API, allowing the coalition to monitor vulnerable flood areas by administrative unit in real time.

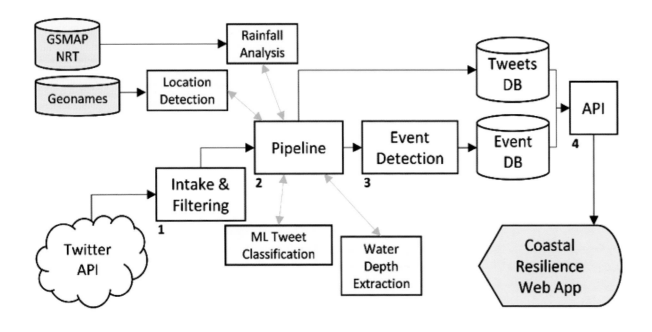

FloodTags performs several analysis steps on the flood related information collected from Twitter, to validate the information, detect locations from it and classify the messages. The analyzed messages as well as the flood events detected from them, are distributed through an API that's connected to the Coastal Resilience web app (preparecenter.org/rcc).

Socio-ecological urban analysis

The socio-ecological analysis supports two workflows within the decision-support tool: flood risk and adaptation solutions. The flood risk portion of the tool helps identify where the most people, places, and livelihoods are at risk from flooding within an administrative unit (kecamatan) and would therefore get the most benefit from nature-based adaptation solutions to reduce this risk. Key flood risk parameters were selected based on availability and scale of the data, summarized within each administrative unit, and spatially joined with the real-time flood reports extracted from FloodTags' API.

The user can view either the most recent flood event or a single selected past flood event, or can specify a date range to view cumulative flood reports. Viewing cumulative reports highlights the most vulnerable areas where nature-based solutions can be prioritized to reduce flood risk. Initially, the summary statistics are provided as a sum across all kecamatans; however, the user can click on an individual kecamatan or administrative unit on the map to visualize the summary statistics in their area of interest. Further, the tool implements the Open Street basemap, which allows users to zoom in to view infrastructure footprints and labels. With this information, a planner or disaster manager can quickly assess potential impacts of a single flood event or highlight areas that show trends of repetitive flooding over time to help prioritize both short-term response and longer-term adaptation actions.

Resilient Coastal Cities Explorer demonstrating tweet-derived floods by the city administrative unit in Semarang, Indonesia.

Theme	Metric per admin unit	Rationale
RISK: Potential flood impacts on people	Number of tweets related to flooding	Real-time social media data on observed flooding helps to ID chronic flooding areas to help prioritize response and recovery actions.
RISK: Potential flood impacts on people	Number of people potentially affected	The more people affected, the more potential socioeconomic impacts felt across the region.
RISK: Potential flood impacts on places	Number of educational buildings potentially affected	Schools and research facilities often provide shelter options during natural hazards. Interruption in educational activities could also have an economic impact if parents are forced to stay home with children.
RISK: Potential flood impacts on places	Number of buildings of worship potentially affected	Places of worship provide shelter and community support during and after natural disasters.
RISK: Potential flood impacts on places	Number of hospitals potentially affected	Could impact care and response time to emergencies.
RISK: Potential flood impacts on livelihoods	Length of roads potentially affected (km)	Flooding of major roads can shut down key routes to economic centers and prevent people from getting to work, negatively impacting their livelihoods and the businesses.
RISK: Potential flood impacts on livelihoods	Area of ag land potentially affected (ha)	Damage to ag land impacts farmers and consumers.
RISK: Potential flood impacts on livelihoods	Number of tourism and culturally important sites potentially affected	These areas provide rec/tourism benefits to the economy.
RISK: Potential flood impacts on livelihoods	Area of rice fields potentially affected (ha)	Damage to ag land impacts farmers and consumers.
ADAPTATION SOLUTION	Number of low production rice fields	Areas that have been deemed "low productivity" by the Ministry of Geospatial Agriculture (Kementrian Pertanian Geospasial) may be areas to prioritize for mangroves/wetland restoration if rice is no longer economically viable.
ADAPTATION SOLUTION	Area of convertible rice fields (ha)	Areas deemed as "rice fields that can be converted" by the Ministry of Geospatial Agriculture (Kementrian Pertanian Geospasial) may be areas to prioritize for mangroves/wetland restoration if rice is no longer economically viable.
ADAPTATION SOLUTION	Area of potentially restorable mangroves (ha)	Historical mangrove areas that have been identified as potentially restorable in prior analyses.

Flood risk and adaptation solution parameters summarized in the Resilient Coastal Cities Explorer decision support tool.

Once a user has explored a community's flood risk, the adaption solution workflow helps visualize where potential nature-based adaptation actions overlap with areas that are reporting flooding based on the selected flood event(s). Adaptation solutions were selected based on actions that are feasible in the region, namely mangrove restoration and open space reclamation/preservation. Key metrics for each parameter were summarized by kecamatans reporting flooding for the selected event(s) and displayed based on solution type with areas that have enabling conditions for more than one adaptation solution highlighted. The ability to view potential adaptation and flood mitigation opportunities alongside real-time flood risk enables planners to consider multiple green or green/gray solutions that will help their communities become more resilient to future flooding while also providing co-benefits back to the community.

Also contributing to the adaptation solutions component are 3D mangrove restoration simulations. The generation of these simulations was inspired by "Building with Nature," a program developed by an organization called Wetlands International. That organization aims to build a stable coastline in Central Java through the construction of permeable dams to support natural mangrove restoration and more sustainable aquaculture production for neighboring coastal communities.

This passive restoration approach uses partially submerged stick structures that allow water to flow through while trapping sediment to mimic the shoreline accretion and erosion control services provided by fully grown mangroves. Once erosion has stopped and the shoreline has accreted (two to five years), mangroves are expected to colonize naturally within three to five years and stabilize the shoreline without manual planting or regular maintenance. As part of the five-year Building with Nature project, the team installed pilot permeable structures in Demak (just northeast of Semarang) in 2015. Since then, they have been tracking progress, and the lessons learned from this case study are intended to be scaled up and leveraged in Semarang in collaboration with TNC Indonesia.

Permeable barrier to trap sediment. (Photo by Nanang Sujana for Wetlands International.)

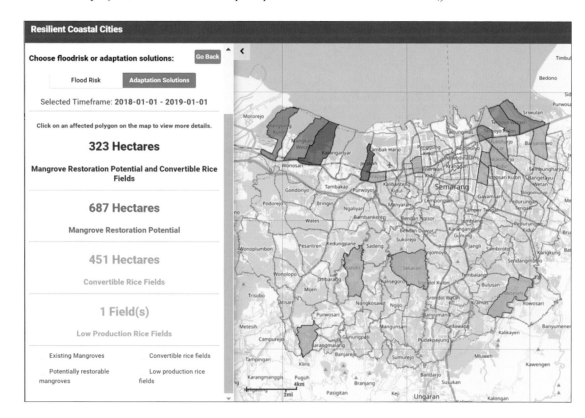

Resilient Coastal Cities Explorer demonstrating possible ecosystem-based adaptation solutions to reduce flood risk by city administrative unit in Semarang, Indonesia.

An ArcGIS Pro model

For a similar approach to be successful in Semarang, the Resilient Coastal Cities team identified the need for city planners to better visualize what a landscape-scale mangrove restoration might look like and how it would grow and provide benefits over time. TNC teamed with Esri to develop an ArcGIS Pro 3D geoprocessing model that simulates permeable dam mangrove restoration scenarios based on known coastal engineering parameters from the Demak case study.

User input parameters included dam location, type, width, deposition depth, elevation from mean sea level, rotation from shore, and sediment accretion bias. Using the Multipatch To Raster tool in ArcGIS Pro, the selected deposition gradients were converted to a raster. The model then compared the output deposition gradient raster to existing bathymetry and, through a conditional statement, found where deposition values are higher than the existing bathymetry and replaced it with the higher value.

For each cell replaced, the model calculated the chance of mangrove growth based on depth, where close to mean sea level means a higher chance for mangrove growth, and no mangrove tree is feasible below a certain depth. A point feature class was created based on the resulting probability of mangrove growth, and its elevation was set to mean sea level. The model then displayed the point feature class as green mangrove trees and the multipatch raster as brown sediment using 3D render classes. The result was a 3D scene layer package depicting sedimentation and mangrove growth behind the dam. The model could then be rerun using different parameters to create scene layers across various scenarios. The output from the mangrove model served two different audiences.

User Inputs

- **Dam location:** Based on clicking and placing dams on map.

- **Dam type:** Concrete, long life permeable, short-term permeable, each with unique representative sediment deposition patterns.

- **Dam width:** In meters.

- **Deposition depth:** Depth of potential sediment accretion (m). At approximately 1 to 3 meters, mangrove creation is greatly affected. Close to mean sea level means a higher chance for mangrove trees; below some depth no tree is possible.

- **Dam elevation:** Height of the dam (m) based on sea level.

- **Dam rotation:** Orientation of dam from shore up to 360° using North as 0°.

- **Sediment accretion bias:** Determines what side of the dam the sediment collects (left, right, center). Based on the water flow and which side of the dam land is present.

Geoprocessing

Create deposition gradient multipatch raster

↓

Compare depth to existing bathymetry and replace as needed

↓

Calculate the mangrove growth chance

↓

Create a mangrove point feature class

↓

Render 3-D styles →

Output

Scene layer package used to either create static mp4 video or published as a hosted web service.

A 3D scene layer package (.slpk)

Workflow of the ArcGIS Pro 3D mangrove restoration geoprocessing model created to simulate sedimentation and mangrove growth due to permeable dam structures. With the interactive model, users can select the placement and characteristics of the dam structure, set sediment accretion rules, and visualize how those parameters might impact large-scale mangrove growth over time.

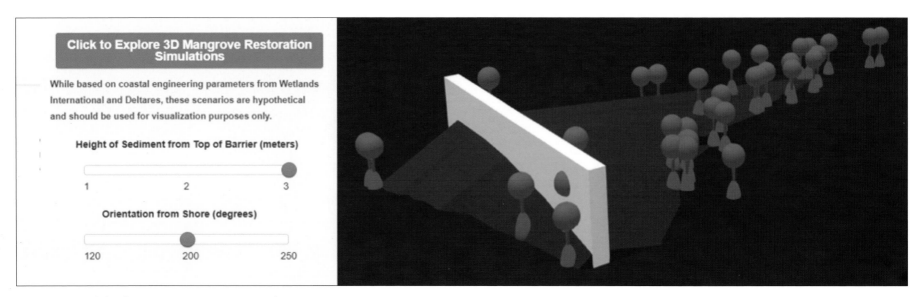

ArcGIS Pro model within Esri Javascript 4x API simulating mangrove restoration scenarios.

For the primary audience of planners and decision makers, a "fly through" style video was created within the ArcGIS Pro model, and then exported to an mp4 viewable in the introduction splash page of the Resilient Coastal Cities app. This capability allows a general audience to virtually observe the benefits that mangrove ecosystems provide to communities over time and influences the audience to consider ecosystem-based adaptation solutions to flood risk. The secondary audience, coastal engineers and scientists, can view scenarios published as web services to TNC's ArcGIS® Enterprise portal, where the scenarios are consumed by an Esri JS 4.x version of the Resilient Coastal Cities app.

This version is currently not publicly accessible, as it is still being adjusted with empirical data collected from pilot projects such as Demak. Once the model is vetted, this version will allow users to assess mangrove restoration project siting, feasibility, and implementation in the Central Java region.

Resilient Coastal Cities Explorer web apps

The Resilient Coastal Cities Explorer web app was built in both 3.25 and 4.0 versions of the ArcGIS® API for JavaScript™ to test the new 3D functionality offered in version 4.0 while still taking advantage of the feature-complete 3.25 version. Both versions incorporate the FloodTags API, which uses flood reports from Twitter returned as JavaScript Object Notation (JSON) data, parsed, and displayed on the web map. Whenever the site is loaded, a query is made to FloodTags for the most recent data, giving the user the feel of a "real time" dashboard.

The 3.25 version of the Explorer app uses a combination of dynamic map service layers and graphic layers to display the data on the web map. The dynamic map service layer consumes vector data from an ArcGIS Server map service and displays it on the map as static reference data. The graphic layers allow user interaction such as hovering and clicking to retrieve information about a specific feature. When a user clicks on the map (graphic layer), a Query function is used. The query returns the graphic's attributes as JSON, which is used to update the data (charts and text) on the app panel.

The 4.0 version of the Explorer app is built differently due to changes within the API that allow 3D data to be displayed on the web map where the user can switch between a scene view (3D) and a map view (2D). An ArcGIS feature layer displays vector data from an ArcGIS Server map service, while a graphics layer leverages a new SimpleRenderer to create 3D graphics by adding a z-value to appropriate points and polygons. The mangrove barrier 3D scenes were created in ArcGIS Pro, published in ArcGIS Online, and consumed in the app with Scene View. Both versions of the web app informed a coalition of primary and secondary audiences to consider the value and critical role that mangroves play in the humanitarian cycle of preparedness, response, and recovery from flooding.

Coalition and community engagement

Integrating the geospatial analysis, model, and simulated mangrove restoration scenarios, we presented the 3.25 version of the Explorer app as the final product to the local Red Cross Society in late October 2018 (preparecenter.org/rcc). The goal of the workshop was to review and solicit feedback on the Resilient Coastal Cities Explorer app to the coalition, with a particular focus on how it could be applied to city planning and coastal management in Semarang and Central Java. For the local Red Cross Society, demonstrating the relationship between coastal flood mapping and the identification of where mangroves could reduce risk in the city provided a fresh perspective to preparedness and recovery planning. Community and stakeholder engagement through tools like FloodTags demonstrated how social media-derived information can highlight the vulnerability of critical infrastructure and illustrate repeat flood areas over time. The Explorer app provides city-specific ecosystem data alongside community assets that can help city level planners and stakeholders make more informed decisions. Presented in relationship to disaster management, appropriate ecosystem-based adaptation solutions became a viable option and a critical part in disaster risk reduction. Workshop attendees were trained on the web-responsive app designed for phones, providing valuable feedback on the application's use for local decision support. A first-of-its-kind ArcGIS Pro mangrove model was built for replication for other coastal cities as part of the greater Resilient Coastal Cities project in Southeast Asia.

Conclusion

The state of the environment lies at the heart of humanitarian action. As coastal development increases and people continue to move to the coasts, we need new and improved strategies and tools for risk mitigation to support growth and minimize stress on the environment. Ecosystems are valuable for mitigating risk and supporting development and humanitarian goals. However, the environment is not systematically considered in global humanitarian action. Failing to integrate the environment within the humanitarian program cycle will further exploit natural resources and limit the safety and prosperity of coastal communities. Only by forming alliances between humanitarian and environmental sectors, as illustrated here with ecosystem-based approaches coupled with geospatial technology, can we achieve viable adaptation solutions in response to coastal climate change.

Through collaborations between TNC and the Red Cross, resilience coupled with geodesign provides critical decision support across communities, nature, and geospatial technology. No longer can we separate our reliance on technology from our place in nature. We must use technology to enhance and promote nature as our communities adapt to ever-changing climate conditions. Our tools for decision support must accommodate relevant information to best respond and meet communities' needs with intuitive design, effective communication, and optimal performance. We must integrate these tools into community engagement and coalition-building to achieve feasible and lasting solutions.

The program of Coastal Resilience has evolved over 11 years of scientific research, tool and web app development, and policy implementation with significant investments in a network of successful conservation and restoration sites. Coastal resilience practitioners have trained and supported more than 100 communities worldwide on the uses and applications of the approach, as well as of the decision-support tool, focusing on the identification of ecosystem-based adaptation and risk mitigation solutions. In partnership with the local Red Cross Society in Grenada, we connected coral reef and seafloor data with wind and wave models to plan for the construction of breakwaters that reduce the threat of coastal erosion while enhancing local fisheries. With accessible science via tools and apps, the community of Grenville Bay moved toward mainstreaming nature-based solutions to increase resilience through restoration in the water. In collaboration with the GDPC, we generated mangrove restoration scenarios to help make important planning decisions for communities faced with flooding and erosion.

Translating complex coastal engineering principles to geospatial models communicated through a web-responsive app provided the City of Semarang with viable solutions to address increasing flood events. The best solutions to climate change and disaster risk reduction may depend less on built infrastructure such as seawalls, groins, and levees, and more on rethinking how we value existing natural resources as part of the equation in achieving coastal protection.

"Impossible is not a fact, it is an attitude."
—Christiana Figueres, Secretariat United Nations Framework Convention on Climate Change

ENDNOTES

Arkema, K., G. Guannel, G. Verutes, S. A. Wood, A. Guerry, M. Ruckelshaus, P. Kareiva, et al. 2013. "Coastal Habitats Shield People and Property from Sea-Level Rise and Storms." *Nature Climate Change*. doi: 10.1038/NCLIMATE1944.

Beck, M. 2014. "Coasts at Risk: An Assessment of Coastal Risks and the Role of Environmental Solutions." A Joint Publication of United Nations University—Institute for Environment and Human Security (UNU-EHS), the Nature Conservancy (TNC), and the Coastal Resources Center (CRC) at the University of Rhode Island Graduate School of Oceanography.

Beck, M., B. Gilmer, Z. Ferdaña, G. Raber, C. Shepard, I. Meliane, J. Stone, et al. 2013. "Increasing Resilience of Human and Natural Communities to Coastal Hazards: Supporting Decisions in New York and Connecticut." *Ecosystems, Livelihoods, and Disaster Risk Reduction*. Bonn, Germany: Partnership for Environment and Disaster Risk Reduction, United Nations University Press.

Beck, M., and C. Shepard. 2012. Environmental Degradation and Disasters. World Risk Report. "Alliance Development Works in Cooperation with United Nations University—Institute for Environment and Human Security (UNU-EHS), and the Nature Conservancy (TNC),"

De Bruijn, J. A., H. De Moel, B. Jongman, J. Wagemaker, and J. C. Aerts. 2018. "TAGGS: Grouping Tweets to Improve Global Geoparsing for Disaster Response." *Journal of Geovisualization and Spatial Analysis* 2, no. 2.

Ferdaña, Z., S. Newkirk, A. W. Whelchel, B. Gilmer, and M. W. Beck. 2010. "Building Interactive Decision Support to Meet Management Objectives for Coastal Conservation and Hazard Mitigation on Long Island, New York, USA." In Building Resilience to Climate Change: *Ecosystem-Based Adaptation and Lessons from the Field*, edited by A. Andrade Pérez, B. Herrera Fernandez, and R. Cazzolla Gatti, 72–87. Gland, Switzerland: IUCN.

Ferrario, F., M. W. Beck, C. Storlazzi, F. Micheli, C. Shepard, and L. Airoldi. 2014. "The Effectiveness of Coral Reefs for Coastal Hazard Risk Reduction and Adaptation." *Nature Communications* 5, no. 3794: doi:10.1038/ncomms4794.

Flessner, L., Z. Ferdaña, G. Guannel, G. Raber, and J. Byrne. 2016. "The Coastal Defense Application: Evaluating Nature's Role in Coastal Protection," In *Ocean Solutions, Earth Solutions*, 2nd ed.. Edited by Dawn J. Wright. Redlands: Esri Press.

Geonames. "About GeoNames." n.d. https://www.geonames.org/about.html. Accessed January 8, 2019.

Groves, C., E. Game, M. Anderson, M. Cross, C. Enquist, Z. Ferdaña, E. Girvetz, et al. 2012. "Incorporating Climate Change into Systematic Conservation Planning." *Biodiversity and Conservation* 0960, no. 3115: 1–21.

Hale, L. Z., I. Meliane, S. Davidson, T. Sandwith, M. W. Beck, J. Hoekstra, M. Spalding, et al. 2009. "Ecosystem-Based Adaptation in Marine and Coastal Ecosystems." *Renewable Resources Journal* 25, no. 4: 21–28.

Hale, L. Z., S Newkirk, and M. W. Beck. 2011. "Helping Coastal Communities Adapt to Climate Change." *Solutions* 2, no. 1. http://www.thesolutionsjournal.com/node/869.

Hürriyetoğlu, A., J. Wagemaker, N. Oostdijk, and A. Van Den Bosch. 2016. "Analysing the Role of Key Term Inflections in Knowledge Discovery on Twitter." Proceedings of the 2nd International Workshop on Knowledge Discovery on the WEB (KDWEB 2016). Cagliari: CEUR Workshop Proceedings.

JAXA. 2018. "JAXA Global Rainfall Watch." https://sharaku.eorc.jaxa.jp/GSMaP/guide.html.

Spalding, M. D., S. Ruffo, C. Lacambra, I. Meliane, L. Z. Hale, C. C. Shepard, and M. W. Beck. 2012. "The Role of Ecosystems in Coastal Protection: Adapting to Climate Change and Coastal Hazards." *Ocean and Coastal Management* 90: 50–57.

Acknowledgments

The work reflected in this chapter could not have been accomplished without the support of current and former TNC colleagues: Nealla Frederick, Boze Hancock and Phil Kramer for their hard work implementing and maintaining the breakwater reef in Grenville Bay, Steve Schill and Lynnette Roth for their GIS work, and Mike Beck, Borja Reguero and Vera Agostini for their marine science expertise, and Ade Rachmi Yuliantri, Glaudy Perdanahardja, Rebecca Scheurer, Jaya Tulha and Jessica Robbins for such great collaboration between TNC and Red Cross in Indonesia. We also want to thank Jurjen Wagemaker at FloodTags for his excellent partnership and product. Many thanks to our colleagues at Esri, including Dawn J. Wright, Shannon McElvaney, Brian Sims, Rich Spencer, Chris Wilkins, and Keith Van Graafeiland for their support of the Grenada work through the production of a collaborative video; and to Eric Wittner and Omar De La Riva for their work on the mangrove ArcGIS Pro model. TNC wants to thank the many partners of Coastal Resilience, which include the National Oceanic and Atmospheric Administration (NOAA), United Nations University, Natural Capital Project, Association of State Floodplain Managers, University of California at Santa Cruz, Critigen, Azavea, Esri, Microsoft, GDPC, and IFRC.

MODELING BIRD RESPONSES TO CLIMATE CHANGE

Using geospatial analysis and mapping tools, a century-old conservation group is targeting which habitats will be most critical for birds in a warmer world, telling stories with maps to show bird lovers just what is at stake and how they can help protect the places that birds and people need to thrive.

By Molly Bennet, with Brooke Bateman, David Curson, Gary Langham, Curtis Smalling, Lotem Taylor, Chad Wilsey, and Joanna Wu, National Audubon Society

The National Audubon Society is committed to helping birds, like this Wilson's warbler, and people weather climate change.

PROTECTING BIRDS YESTERDAY AND TODAY

A conservation organization founded in 1896 is tackling a modern-day problem

More than 700 species of birds make their home in North America. And for more than a century, Audubon's mission has been to protect them all. That mission began, surprisingly, with a fashion trend—specifically, the elaborate hats adorned with magnificent waterbird feathers that were once *de rigueur* for stylish women in North American cities.

By the late nineteenth century, the demand for plumed hats was so strong that it threatened to extirpate species such as the great egret, whose delicate white feathers were worth twice their weight in gold. In 1896, two Bostonian women, Harriet Hemenway and Minna Hall, decided to save the species from the fashion industry, so they founded the Massachusetts Audubon Society. They named the nascent conservation group after ornithologist and artist John James Audubon, and their goal was simple: persuade society ladies to give up their plumed hats. During the next decade, other groups founded Audubon societies, and these groups eventually joined together in 1905 as the National Audubon Society. All were motivated by a single idea: Birds are something to be protected, not exploited.

As it grew over the course of the twentieth century, Audubon pursued that mission from multiple angles. A constellation of Audubon-managed sanctuaries has provided refuge for birds as development swallowed up avian habitats. During its history, Aububon has successfully advocated for policy changes, such as the 1918 Migratory Bird Treaty Act, which granted federal protections to all migratory birds, and, later, the ban on the eagle-killing insecticide DDT. Audubon now has more than 450 local chapters across the country.

Today, however, more than a century after early conservationists saved species such as the great egret from the ravages of the plume trade, birds face a new and unprecedented global crisis caused by the ecological impacts of climate change, an impending catastrophe that is global in scale and yet also intensely local in its effects.

Audubon scientists have a two-fold task to protect birds from this existential threat. First, they're working to understand the nature of the threat itself—predicting how and where climate change will affect individual species in the next decade. Second, they're using predictions of where different bird populations might move as habitats change. These predictions can inform conservation work, identifying where conservationists can work strategically to protect those places that will provide havens for birds in a warming world.

Mapping programs such as ArcGIS have played an invaluable role in these projects, from identifying coastal wetlands that could serve as lifelines for salt-marsh species as sea levels rise to helping community scientists document local birds' response to climate change in real time. Audubon scientists in turn have used maps to communicate the science to a wider audience, translating the complex relationships between birds and climate into an intuitive visual language. This work allows bird lovers everywhere to fully grasp an abstract phenomenon such as climate change, showing how it will impact the birds in their own backyards. These efforts make a persuasive case for a strong conservation ethos among the public and elected leaders so that the birds we know and love have a fighting chance to adapt in a changing world.

In the late nineteenth century, women's hats were adorned with feathers, the collection of which threatened the extinction of several bird species. Facing page, a painting of a snowy egret by John James Audubon.

CLIMATE 1.0—MAPPING WHERE THE BIRDS WILL GO

Scientists for decades have gathered evidence that climate change is profoundly affecting wildlife. But they still must answer a complex question: How will birds react to habitat changes in precipitation and temperature? Will bird populations adapt in place, move to habitats with more favorable conditions, or simply struggle and even disappear? For conservationists, answering that question is critical to preparing for the future. "You can't save what you don't understand," says Gary Langham, Audubon's chief scientist. "If we don't know which birds are most vulnerable, it's hard to take action to protect them." In the late 2000s, Langham and his team set out to identify that threat. To do so, Audubon's scientists and statisticians needed to analyze the climate preferences of hundreds of different species of North American birds, and then determine whether and where those characteristics might be found as the climate changes.

To determine those preferences, Audubon relied on its Christmas Bird Count community science program and 117 years of longitudinal data showing wintertime species occurrence across the United States and Canada. By analyzing Christmas Bird Count data from 2000 to 2009 against climate records from the same period, Langham and his colleagues determined the preferred temperature and precipitation characteristics—known as "climate suitability"—for hundreds of North American bird species in their wintering grounds. Data from the Breeding Bird Survey, a similar bird-counting census that takes place every summer, identified climate suitability for those same species during their breeding periods.

The next step was to look into the future. Langham and his colleagues used projections from the Intergovernmental Panel on Climate Change on how North American climate conditions would likely change based on different carbon emissions scenarios. Then they chose a resolution of 10-by-10-kilometer squares to map North American geographic areas likely to provide suitable climate conditions for each bird species. Those maps revealed where a species such as the common loon might find hospitable breeding grounds in the future. The maps also show the degree to which those grounds overlap with the bird's current range and whether it may be forced to shift its range to follow climate conditions. Crucially, the maps identify how much area remains that meets the climate requirements of the species. The information shows which species might suffer the most from climate change.

Audubon's 2014 *Birds and Climate Change Report* released the ominous results. Of the 588 bird species analyzed, 314 could find their ranges shrinking by more than 50 percent by 2080, under even the most optimistic emissions scenario. Some of those birds, like the common loon, may see suitable climatic conditions open up outside their current range, meaning that by the end of the century, while Minnesotans may no longer hear the loon's call in the summer, the species may find refuge farther north, in Canada. For others, like the Baird's sparrow, projections do not show any new habitat to move to.

Despite being grim, the findings illuminated a path forward. Climate conditions are a key factor in whether a given geography may be suitable for a species to overwinter or breed. However, a complex web of ecological variables determines whether a place will actually meet that species' needs. (Audubon's scientists have already begun incorporating some of those variables—such as vegetation cover and land use—into its models as part of an ongoing Climate 2.0 project.) For more than a century, Audubon has worked to protect and manage critical habitats so that bird species can thrive. The climate range maps produced in Audubon's study didn't just reveal which species will be most vulnerable to climate change; they also highlighted the geographies that could be lifelines for those species as the climate changes—those places where conservation work and habitat protection will prove most critical now and in the future.

To unravel the mystery of where bird species will go as the climate changes, Audubon's scientists looked to the past.

COMMON LOON

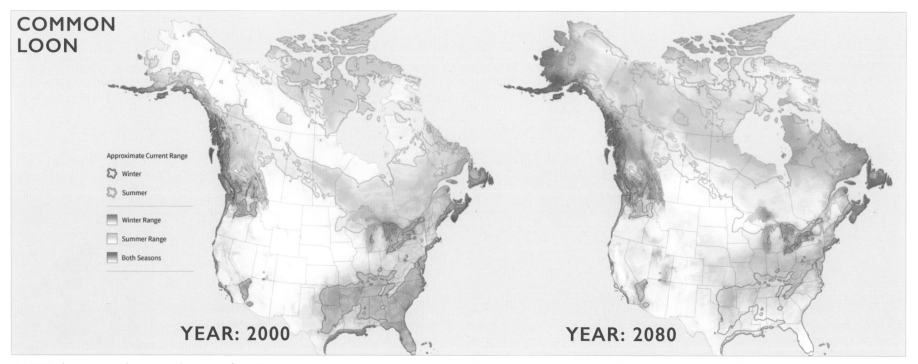

Approximate Current Range

⬡ Winter

⬡ Summer

▢ Winter Range

▢ Summer Range

▢ Both Seasons

YEAR: 2000

YEAR: 2080

By 2080, the common loon may disappear from Minnesota as its summer range moves north, according to Audubon's Climate 1.0 projections. The species' winter range could also be heavily affected, declining 62 percent by 2050. The species has already disappeared from some former nesting areas because of human disturbance on lakes in summer. Acid rain may also reduce food supplies in the loon's breeding range.

Common loon

The common loon has been protected on some breeding grounds in the Northeast by volunteer "Loon Rangers" who patrol the lakes and help educate the public about conservation. Projections show the loon will lose much of its breeding range as a result of climate change.

BOBOLINK

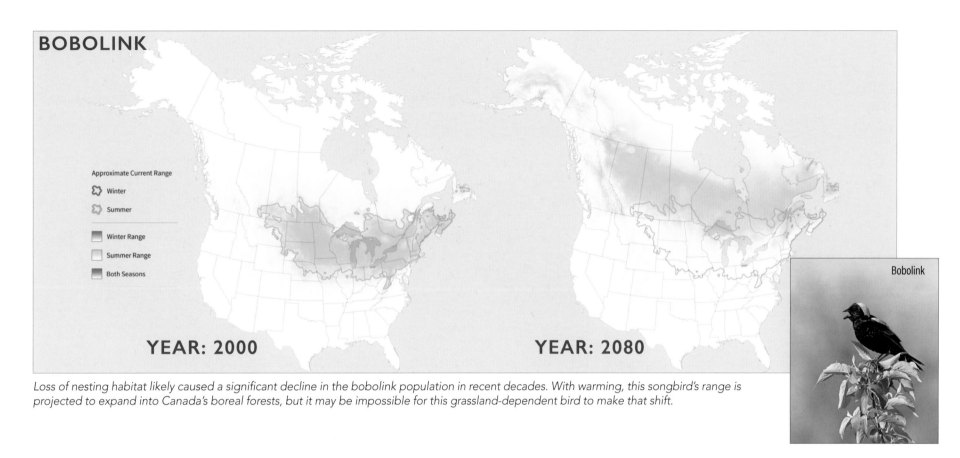

Approximate Current Range

⬡ Winter

⬡ Summer

⬛ Winter Range

⬜ Summer Range

⬛ Both Seasons

YEAR: 2000

YEAR: 2080

Bobolink

Loss of nesting habitat likely caused a significant decline in the bobolink population in recent decades. With warming, this songbird's range is projected to expand into Canada's boreal forests, but it may be impossible for this grassland-dependent bird to make that shift.

EUROPEAN STARLING

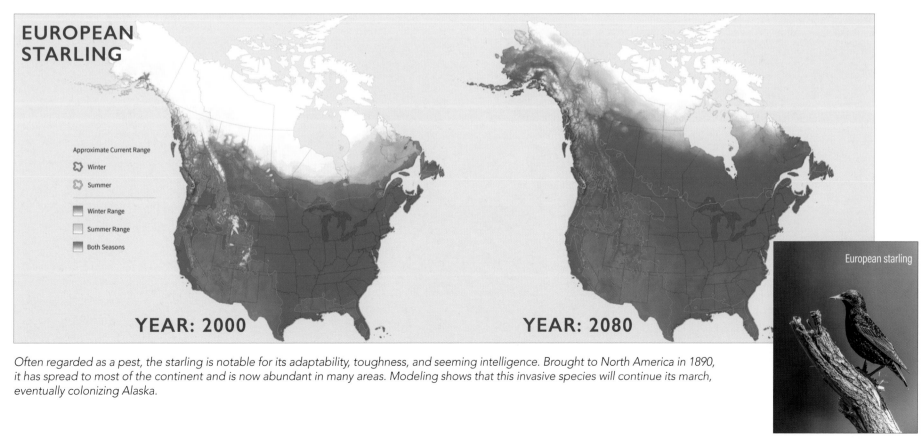

Approximate Current Range

⬡ Winter

⬡ Summer

⬛ Winter Range

⬜ Summer Range

⬛ Both Seasons

YEAR: 2000

YEAR: 2080

European starling

Often regarded as a pest, the starling is notable for its adaptability, toughness, and seeming intelligence. Brought to North America in 1890, it has spread to most of the continent and is now abundant in many areas. Modeling shows that this invasive species will continue its march, eventually colonizing Alaska.

CASE STUDY: NORTH CAROLINA
Mapping where climate strongholds will be

A team led by Curtis Smalling, the director of conservation for Audubon North Carolina, has been building on the analyses of Audubon's *Birds and Climate Change Report*, which mapped where different bird species may find the needed climate conditions to survive. To determine which regions of the state could serve as "climate strongholds," Smalling and his colleagues grouped the list of climate-threatened species in the state into four guilds on the basis of their habitat needs—coniferous forest, deciduous forest, open agricultural lands, or high elevation—and analyzed projections of where each species could find suitable climatic conditions both today and in the future. Audubon's scientists already knew which North Carolina locations provided critical habitat to those bird species. That knowledge was based on previous work that identified Important Bird Areas (IBAs)—a global project initiated by BirdLife International to find places that are crucial to birds. Overlaying the climate data with the IBA maps gave Smalling and his colleagues insight into which of those places will increase in importance as the climate changes. "Eighty years from now, are they still going to be the best places for these birds?" asks Smalling. "Those places that hit on all cylinders rise in our thinking."

One such place is the Blue Ridge Mountains in western North Carolina. At 1.1 billion years of age and counting, the range is geologically one of the oldest in the world. The range provides distinct habitats, from deciduous forest to open agricultural land, that support birds such as the chimney swift and the golden-winged warbler. As the climate changes, Audubon's models suggest that the cooler, wetter climate found in the upper elevations of the range could provide sanctuary for birds whose current homes are becoming warmer and drier. Audubon works to protect land in these areas from development and helps landowners increase the diversity of habitat on their properties—appealing to a greater number of bird species—with the understanding that this landscape will become more important as the climate changes. "It is a powerful argument for land protection and land management," says Smalling.

The Blue Ridge Mountains—named for their bluish color when seen from a distance—make up an important climate stronghold for endangered birds.

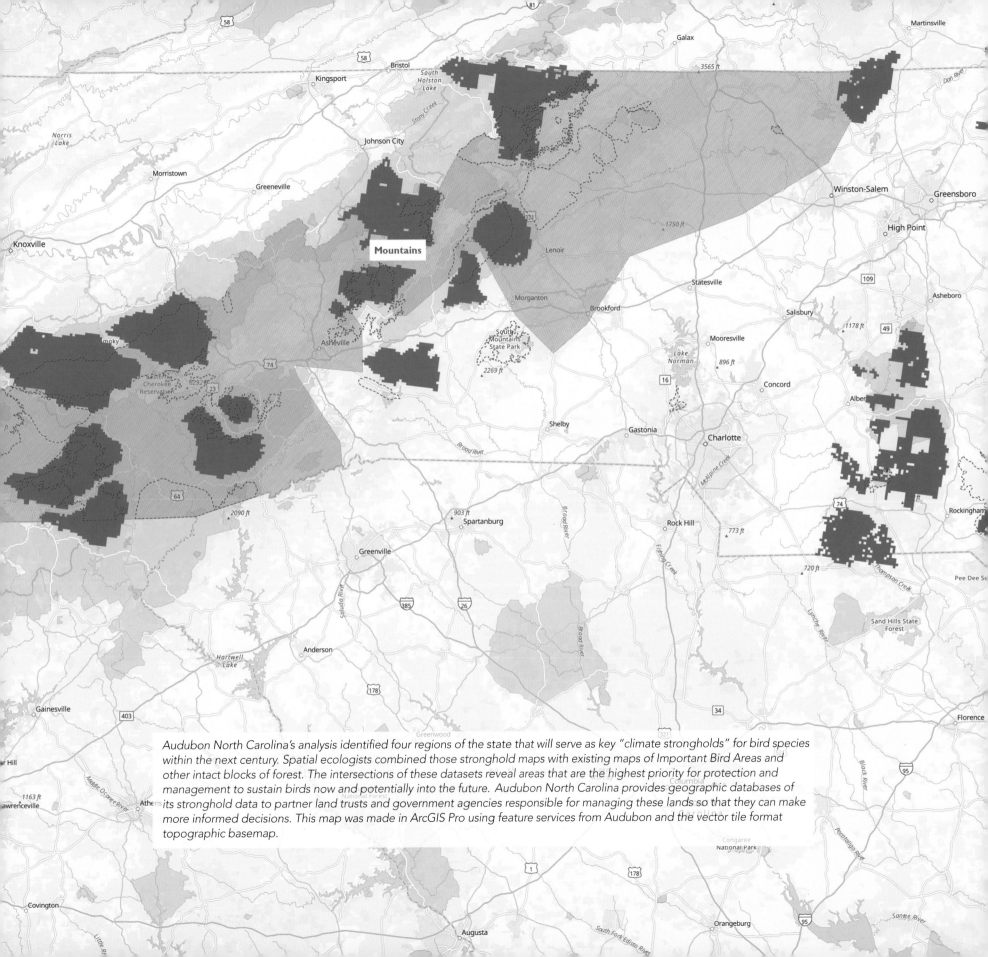

Audubon North Carolina's analysis identified four regions of the state that will serve as key "climate strongholds" for bird species within the next century. Spatial ecologists combined those stronghold maps with existing maps of Important Bird Areas and other intact blocks of forest. The intersections of these datasets reveal areas that are the highest priority for protection and management to sustain birds now and potentially into the future. Audubon North Carolina provides geographic databases of its stronghold data to partner land trusts and government agencies responsible for managing these lands so that they can make more informed decisions. This map was made in ArcGIS Pro using feature services from Audubon and the vector tile format topographic basemap.

Roanoke and
Chowan Rivers
Bottomlands

Capital-Piedmont

Southern
Coastal
Plain

Important Bird Areas (IBAs)

Priority Forest Blocks

Climate Strongholds

20
Miles

BIRDS WITHOUT BORDERS

Protecting species on both sides of their migration

People have known for decades that many bird species travel thousands of miles between their wintering and summering grounds. Scientists had believed that a species' ability to thrive was driven in large part by the conditions of the summer breeding habitat. But new research that tracks birds as they migrate shows that the conditions birds face in their winter homes and on their migration routes also affect birds' long-term population numbers. Scientists now understand that rising sea levels, increasing storms, and warming temperatures in wintering grounds such as Colombia and the Bahamas can threaten the birds seen in, for example, backyards in northern Michigan on a summer day. One of Audubon's priorities in recent years has been to identify locations in Latin America that will be most important to birds as those changes begin to be felt.

To accomplish that task, Audubon's International Alliances Program worked with Chad Wilsey, vice president of conservation science, and partners from BirdLife International and Durham University to explore how species distributions are likely to change in nearly 1,200 IBAs in 12 countries. By analyzing current and projected climate conditions in those IBAs alongside existing climate vulnerability assessments for nearly 4,000 bird species, they determined the degree of turnover each IBA is likely to see—how many species will potentially disappear (known as "extirpation") as the climate changes, and how many new species will potentially appear ("colonization"). From there, they grouped each IBA into one of five different adaptation-management classes based on the anticipated change in number and types of species; IBAs that are expected to retain their current species distributions or even increase the number of species they host could be good places to focus conservation efforts and resources or to advocate for increased government protections. "This scientific analysis provided a backbone for developing climate change adaptation strategies for partner countries," says Wilsey. "Those plans are now influencing conservation and policy decisions across the region."

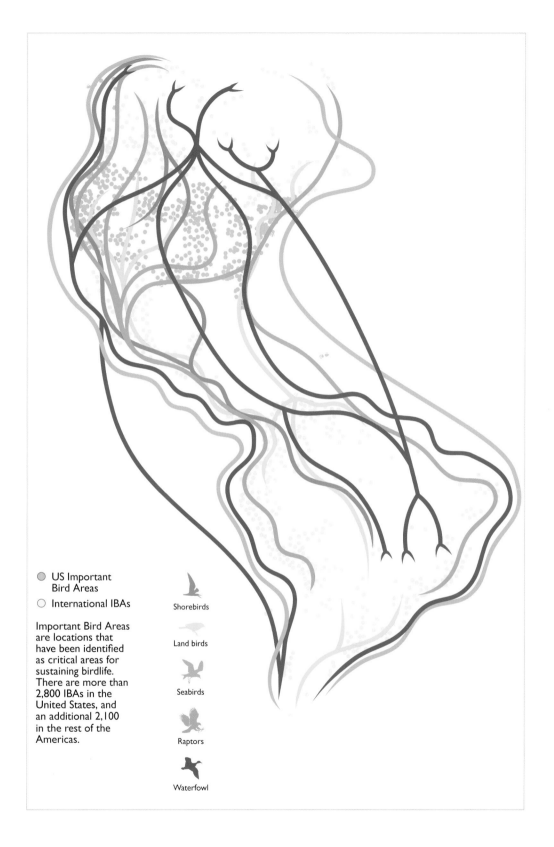

○ US Important Bird Areas

○ International IBAs

Important Bird Areas are locations that have been identified as critical areas for sustaining birdlife. There are more than 2,800 IBAs in the United States, and an additional 2,100 in the rest of the Americas.

Shorebirds

Land birds

Seabirds

Raptors

Waterfowl

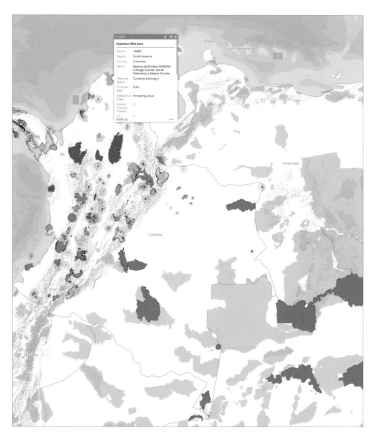

Colombia (left) is the winter home for many songbird and raptor species that breed in the United States, including the golden-winged warbler. As part of its analysis, Audubon examined what climate change might mean for bird species in three of Colombia's national parks: Parque Nacional Natural Chingaza, Santuario de Fauna y Flora Guanentá–Alto Río Fonce, and Parque Nacional Natural Serranía de los Yariguíes.

Sandhill cranes fly over a wetland in the state of Chihuahua, Mexico, while a local farmer looks on.

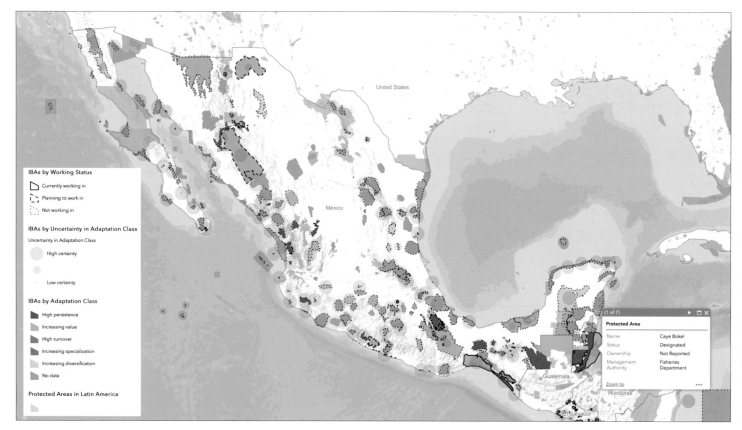

Audubon's analysis of how species distribution will change in Latin America found that Mexico is a hotbed for areas that could both sustain the species that currently live there and see new species arriving in search of suitable climate. In the northern part of the country, the Chihuahuan Desert, which straddles the US–Mexico border, may provide a haven for climate-threatened grassland species. Audubon's scientists are mapping habitat and land use in that region as part of Climate 2.0 to help inform conservation priorities.

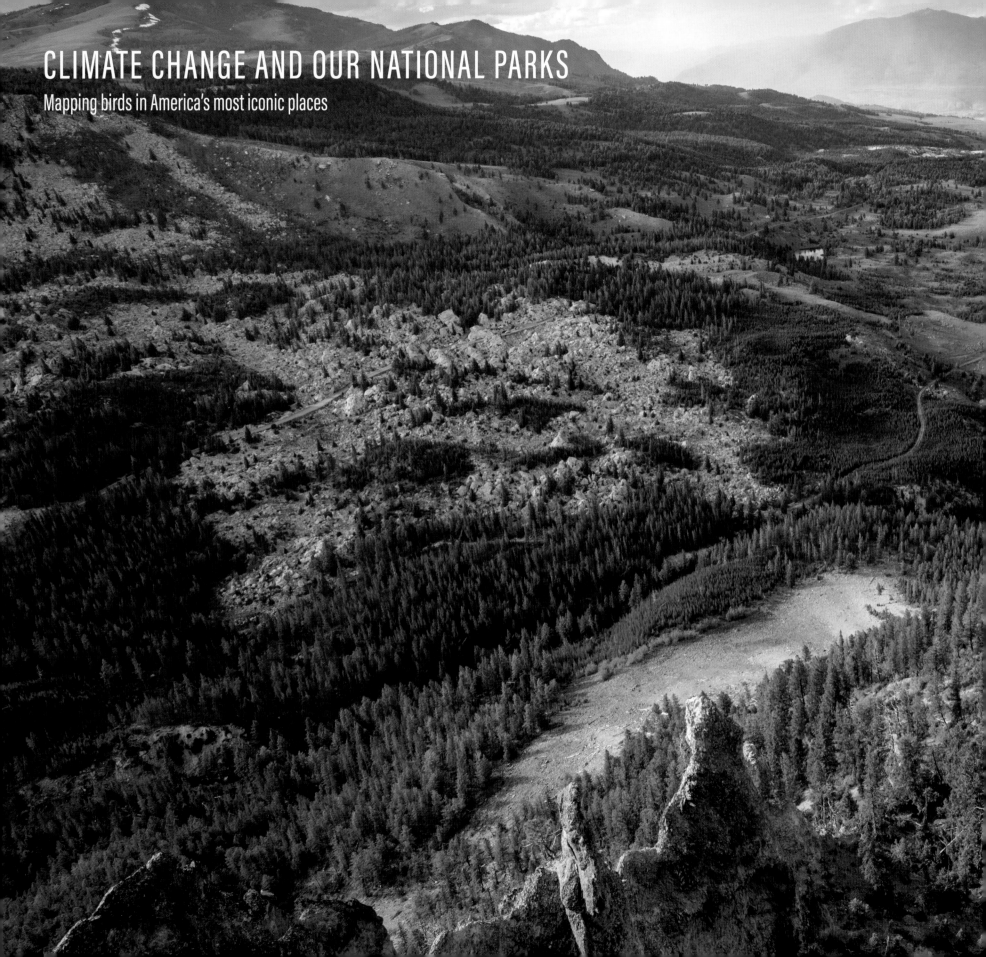

CLIMATE CHANGE AND OUR NATIONAL PARKS

Mapping birds in America's most iconic places

Congress created the National Park Service in 1916 to manage and protect the country's most valuable ecological and cultural resources. The step was a remarkable act of foresight. As the world around them has been built up and developed during the last century, the 84 million acres of public land that make up the national park system have nurtured hundreds of bird species, from the riparian songbirds that nest among the riverside willows of Wyoming's Yellowstone National Park (shown here) to the warblers that summer in the boreal forest of Acadia National Park in Maine. But as the climate changes, those diverse ecosystems will also change, making long-term, strategic conservation planning in each park critically important both to the birds that depend on the parks now and those that may seek refuge in the future.

To understand the role our national parks will play in the survival of different bird species, Audubon and the National Park Service (NPS) set out to study how the distribution of species across the park system may shift over the next three decades—which parks will lose birds, which will gain them, and which may see a wholesale turnover of species. For each of 274 sites managed by the NPS, the team gathered lists of bird species that are currently known to spend summers or winters in a given park. The team then overlaid data from the Audubon *Birds and Climate Change Report*, which showed where those species are likely to find suitable climate in the year 2050 if global carbon emissions continue apace or if they're sharply reduced.

In parks the team studied, a quarter of the bird species on average will change by 2050 unless emissions are reined in. Turnover rates are predicted to be particularly high in the Midwest and Northeast, which may see an exodus of boreal-forest breeding birds heading toward Canada and an influx of new species seeking refuge from warming temperatures in the southern United States. In some cases, a species that currently makes its summer or winter home in a given park may decline or potentially leave the park altogether by midcentury. For example, the bald eagle could potentially disappear from Grand Canyon National Park. Overall, however, the researchers found that the parks will actually gain more species than they lose, becoming critical refuges for species forced to shift their range as they seek suitable climate conditions. Knowing which species may struggle and which newcomers may arrive will help NPS ecologists manage the habitats under their care.

Wilson's warbler

Bald eagle

The bald eagle could potentially disappear from Grand Canyon National Park.

Winter turnover

In winter, all parks will likely experience more colonizations and extirpations, which together amount to more turnover. As birds relocate from the South, they may concentrate in the Midwest and Northeast.

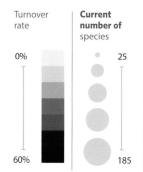

Reading these maps

Each circle represents one of the 274 parks analyzed for climate–driven changes in the ranges of North American bird species. Summer population distributions are modeled from North American Breeding Bird Survey data; winter distributions are modeled using data from the Audubon Christmas Bird Count.

Turnover rate

0%

60%

Current number of species

25

185

Summer turnover

Parks in the Midwest, Northeast, and the Rockies are expected to have the highest species turnover in summer. The Rockies especially will see a wave of dry-forest birds and a departure in alpine-loving birds.

Climate is just one factor in where a species ends up. Future ranges will also be influenced by availability of food, water, and shelter. And there's the question of whether birds will travel to places that match their pre-ferred conditions, or simply adapt to their surroundings.

While this report only looks at precipitation and temperature, other climatic impacts like sea-level rise, wildfires, stronger hurricanes, earlier snowmelt, and invasive species are already affecting avian distribution and habitat.

Certain seabirds and species often missed or overlooked in terrestrial bird counts aren't included in the analysis.

The analysis probably underestimates turnover in southern parks, as it doesn't include Central and South American species, which could move into the United States in coming decades.

These maps depict an analysis of how species distribution might change within the boundaries of 274 national park units in summer and winter months, based on projections from Audubon's climate models. The maps indicate which parks are likely to see a high degree of extirpation— in other words, bird species that currently inhabit the park leaving as the climate becomes unsuitable for them—and which will see a high degree of colonization by new species.

CASE STUDY: MARYLAND

Engineering climate resilience in the Chesapeake Bay

The tidal marshes of the Chesapeake Bay are renowned for the fragile ecosystems they nurture, providing critical nesting ground for black rail, saltmarsh sparrow, and other species, and vital nurturing grounds for economically important fisheries. But with sea levels in the Chesapeake Bay expected to rise by three to five feet by the end of the century, much of the habitat these birds and fish depend on will soon be underwater. Preparing upslope waterlogged habitats as functional tidal marshes is one way to combat the effects of sea-level rise.

"As we go forward through this century, we're going to lose a lot of tidal marshes," says David Curson, director of bird conservation for Audubon Maryland–DC. Populations already in decline, such as the black rail and saltmarsh sparrow, will acutely feel the loss. "One component of Audubon's climate-adaptation strategy for tidal marshes is to compensate for that marsh erosion by facilitating new marsh being formed on adjacent uplands as tides reach further upslope and making sure that new marsh is suitable for salt marsh–dependent bird species," he says.

This process, called "marsh migration," begins with salt water seeping into inland soil and killing forest vegetation, allowing marsh grasses to grow in its place. The death of the trees, however, can sometimes cause the ground surface to drop a couple of inches, forming a shallow basin that retains floodwater and leaves the transitioning marsh waterlogged and unsuitable for birds such as the saltmarsh sparrow. And worse, the waterlogged conditions weaken the new marsh vegetation, resulting in erosion into open water, jeopardizing the entire transition process.

In the spring of 2018, Audubon Maryland–DC and a handful of conservation partners began an engineering project aimed at enhancing necessary tidal flow to one such site, Farm Creek Marsh, by cutting a channel that will link the marsh to a tidal creek 300 yards away—giving floodwaters a place to drain.

As the Farm Creek Marsh project proceeds, Audubon scientists are working to identify other transitioning marsh sites throughout the Chesapeake Bay. These sites, like Farm Creek Marsh, could serve as climate change sanctuaries if their hydrological function is restored. To find these sites, Audubon turned to spatial datasets, such as the National Wetlands Inventory and the European Space Agency's global surface water data, to determine whether any characteristics found in those datasets corresponded with one of the qualities that makes a marsh a good candidate for this kind of restoration. These qualities include estuarine vegetation, the presence of standing surface water as evidence of waterlogging, and the absence of any connection to a tidal creek. When overlaid, the maps overlapped in more than 200 distinct places, indicating key candidates for further investigation by Curson, who will visit potential sites in the summer of 2019.

How Audubon scientists used geospatial analysis to identify the most functional future marsh and wetland areas in the Chesapeake and Delaware Bays

Saltmarsh sparrow

Lotem Taylor, a GIS and data analyst for Audubon, used the National Wetlands Inventory, a nationwide inventory of wetlands identified through aerial imagery analysis, to identify present-day high marshes in the region. Using that dataset and sea-level rise projections from NOAA, Taylor and her colleagues identified sites suitable to remain marsh through 2050.

As an indicator of marsh degradation, Taylor looked for sites where standing surface water increased during the past few decades, a change that occurs when marshes cannot keep pace with sea-level rise. She identified transitions toward an increase in water from the Global Surface Water Dataset, a project that uses Landsat satellite imagery to map the spatial and temporal distribution of surface water since 1984 at a 30-meter resolution.

To assess drainage ability, Taylor and her colleagues analyzed the distance to tidal creeks and underlying soil composition using the Soil Survey Geographic Database, the National Hydrography Dataset, and National Wetlands Inventory.

To identify suitable sites for restoration, Taylor used ArcGIS Pro to run a spatial overlay analysis. She clipped current marsh by areas of surface water increase to identify waterlogged sites, and then aggregated the resulting polygons to group features that were small and close together, using a minimum size of 0.5 hectares and a distance of 30 meters to match the mapping resolution. She kept polygons that intersected marsh in the future to include sites where restoration could have a lasting impact. Finally, Taylor kept sites with poorly drained soils that were disconnected from tidal creeks.

This analysis identified 239 sites covering approximately 275 hectares in the Mid-Atlantic that could potentially benefit from restoration. The majority of these sites sit along the Chesapeake Bay's eastern shore. Dorchester County, Maryland, accounted for more than a third of the output in terms of number of sites (110) and area (139.9 hectares). Coastal New Jersey also had a high number of sites (57, 64.6 hectares) in Ocean, Cape May, and Atlantic Counties. Few sites were identified along the western shores of the Chesapeake and Delaware Bays and none on the upper Chesapeake Bay.

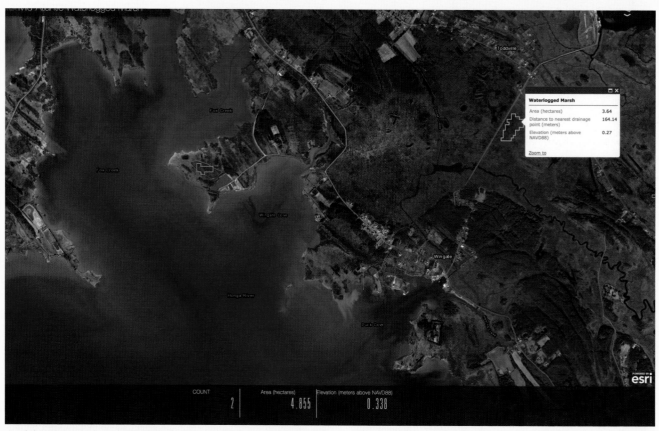

Audubon's spatial analysis of wetlands in the Chesapeake Bay pinpointed more than 200 sites that have the potential to serve as refuges for saltwater marsh birds as rising sea levels engulf much of the current bird habitat.

CLIMATE WATCH
Using community science to validate climate models

Climate Watch volunteers observe and record what species they see and hear to help refine Audubon's climate models.

For bird lovers who survey their communities during the Christmas Bird Count in December or await the arrival of warblers in their local parks every spring, climate change isn't some abstract problem looming on the horizon. Rather, it's a personal one whose effects are already felt on the birds they've come to care about. "Climate change is happening everywhere, it's global," says Brooke Bateman, senior climate scientist at Audubon. "But at the same time, it's so local—it's different everywhere." In 2016, Audubon piloted a new project called Climate Watch in which hundreds of volunteers help Audubon track how birds react to climate change in their own communities in real time—and gather data that's helping Audubon's scientists test and refine their climate models.

Like the Christmas Bird Count, Climate Watch is essentially a survey: volunteers record the birds they see and hear. They aim to answer a specific question: Are particular species (the project is now focusing on several bluebird and nuthatch species) abandoning or colonizing a given area as the climate-suitability models predicted?

To validate Audubon's climate models, surveys must follow a rigorously structured protocol; ArcGIS has been a valuable tool for coordinating them. Using Audubon's Esri Climate Watch Planner web map and Claim a Climate Watch Square mapping tool, a volunteer selects one of the seven target species they'd like to survey and the season (the surveys take place twice a year, during the winter and summer), and then claims a 10-by-10-kilometer "Climate Watch Square." The volunteer sees how the climate suitability in that square is expected to change for the targeted species based on Audubon's models and plots 12 locations they'll survey within that square, choosing spots where the birds are most likely to be.

Audubon's scientists crunch the data collected during the surveys using occupancy-estimation models that assess the likely presence or absence of a species within each Climate Watch square. But to determine whether any eastern bluebirds are hiding in those 100 square kilometers, volunteers must look in the right places. "We really depend on local knowledge," says Bateman. In cases where populations of a target species are expected to decline, that means surveying locations that the birds have used consistently in the past. Or if a species is expected to colonize an area, it means knowing where to find the habitat they'd gravitate toward. Once volunteers select the 12 survey points for each selected square, they return to the same spots each year. "That's how we're going to be able to detect change," says Bateman. "It's the same location, the same habitat. The only thing that's changing is the climate."

Beyond the scientific value of the Climate Watch surveys, Bateman hopes that engaging everyday bird lovers in observing the real-time effect of climate change in their communities is empowering. "When you see things change, you become the expert on your local responses to climate change," says Bateman. "You can go to your elected officials and say, 'This is happening, and I'm seeing it with my own eyes.'"

Brown-headed nuthatch

A birdwatcher's tools

ENGAGING VOLUNTEERS TO HELP TRACK CLIMATE CHANGE

Climate Watch aims to document birds' responses to climate change. Volunteer community scientists look for bluebirds and nuthatches in the places where Audubon's climate models project they should or shouldn't be in the next decade.

The Climate Watch team built interactive tools on the Esri platform so that volunteers could identify and select areas in their region that will be most helpful to support the survey efforts.

In this map for the Eastern bluebird, Climate Watch squares in Madison, Wisconsin, show where this species is predicted to see improving summer climate conditions (orange) or no change in climate conditions (gray). Each blue dot, which is identified by local volunteers as areas where one would expect to see Eastern bluebirds, represents a survey point within the 10-by-10-kilometer square.

Because Audubon's Climate Watch happens twice a year, volunteers must specify when they're claiming a square, which species they're studying, and if they're claiming the square for summer, winter, or both. Once volunteers claim their squares, they're sent back to the Climate Watch planning tool to plot their survey points.

For now, data delivery to the Climate Watch team happens over email. But in the future, the Climate Watch team hopes that volunteers can log their sightings automatically using Audubon's updated Bird Guide mobile app.

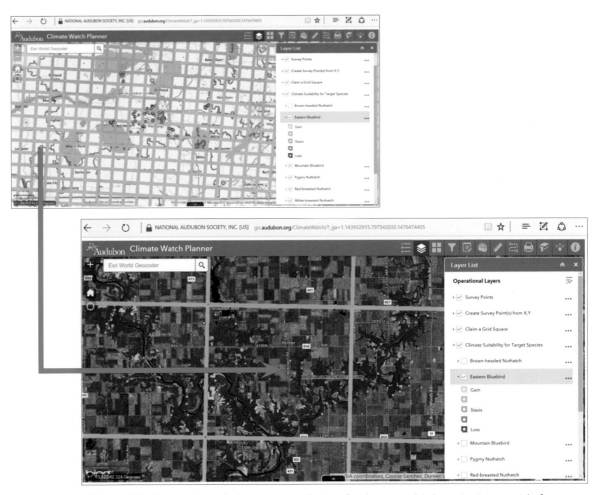

Climate Watch scientists split the range predictions for the target bird species into a grid of 10-by-10-km Climate Watch Squares that cover the United States.

Once volunteers decide which bird they want to study, and where they want to do their surveys, they're directed to the Claim a Climate Watch Square Esri tool to get started.

WHY BLUEBIRDS AND NUTHATCHES?

It would be nearly impossible to complete thorough and timely surveys for each of the hundreds of climate-threatened species identified in Audubon's *Birds and Climate Change Report*. So during its first three years, Climate Watch focused on seven ideal species for study based on a few important factors. First, Audubon's scientists knew that the range and climate-suitability data for these species was highly accurate. All the species are predicted to see significant range shift, making them strong test cases for model accuracy. These species are also relatively easy for nonexpert volunteers to identify during surveys, making the project accessible to almost anyone who wants to participate. Many people also like bluebirds and nuthatches and care about what's going to happen to them. The survey added orioles and robins in 2019.

Red-breasted nuthatch

Mountain bluebird

Pygmy nuthatch

ENDNOTES

Pages 118–19: Wilson's warbler. Photo: Camilla Cerea/Audubon.
Page 120: Snowy egret. Illustration: John James Audubon.
Page 121: Hats. Photo: National Audubon Society.
Page 122: Drought. Photo: Joseph Helfenberger/Adobe Stock.
Page 123: Common loon. Photo: Brian Lasenby/Adobe Stock.
Page 124: Bobolink. Photo: M. Truchon/Adobe Stock.
Page 124: European starling. Photo: Randy Anderson/Adobe Stock.
Page 125: Blue Ridge Mountains. Photo: Camilla Cerea/Audubon.
Page 129: Beach. Photo: Camilla Cerea/Audubon.
Pages 130–31: Yellowstone National Park. Photo: Jacob W. Frank/NPS.
Page 132: Wilson's warbler. Photo: Design Pics Inc./Alamy.

Page 132: Bald eagle. Photo: Andy Morffew.
Page 133: Maps by Stamen Design.
Page 134: Chesapeake. Photo: Camilla Cerea/Audubon.
Page 135: Saltmarsh sparrow. Photo: 500px/Alamy.
Page 136: Volunteers. Photo: Camilla Cerea/Audubon.
Page 137: Brown-headed nuthatch. Photo: Matt Cuda/Adobe Stock.
Page 137: Maps and bincoculars. Photo: Camilla Cerea/Audubon.
Page 139: Mountain bluebird. Photo: Rck/Adobe Stock.
Page 139: Pygmy nuthatch. Photo: 500px/Alamy.
Page 139: Red-breasted nuthatch. Photo: Steve Byland/Adobe Stock.

PART 3
HOW WE LOOK AT EARTH

Successfully understanding how Earth works and how Earth looks to us requires integrative and innovative approaches to observation and measurement. These approaches include Earth observation in varying forms, such as from sensors on satellites, aircraft, drones, ships, and so on. They also include the important data science issues of conducting analysis; modeling, developing, and documenting useful datasets for science; and interoperating between these datasets and between various approaches.

Updated every 16 days for any given location, the Normalized Difference Moisture Index (NDMI) estimates levels of moisture in vegetation. Wetlands and other vegetated areas with high levels of moisture appear as blue whereas deserts appear as tan to brown.

MAPPING ANCIENT LANDSCAPES

Racing against the clock as development encroaches on important Kurdish heritage sites, a team of landscape archaeologists deploys drones and comparative image analysis to capture previously undetected ancient settlements.

By Jason Ur and Jeffrey Blossom, Harvard University

Drone-acquired image[1] of the site called Qasr Shamamok, located not far from Erbil, in the Kurdish region of Iraq. It was recently excavated by a team of archaeologists who showed that the occupation probably dates to the Chalcolithic, with important Assyrian remains from the late fourteenth century to the seventh century BCE.

THE ASSYRIAN EMPIRE

In late spring of the year 612 BCE, the combined armies of the Babylonians and Medes arrived outside the gates of Nineveh on the Tigris River in what is today northern Iraq. At the time, Nineveh was still the capital of the Assyrian state, but Assyria's fortunes were on the wane. Two years earlier, the ancestral city of Ashur had fallen to the Medes. The attacking forces besieged the city for a few months before ultimately breaking through its defenses and sacking and burning it. The royal court survived for a few years in northern Syria, but for all intents and purposes, the Assyrian Empire came to a violent end that day. It was an ignominious end for what had been the most powerful state in the Near East in its day and arguably the first unambiguous world empire.

At its height, the Assyrian kings ruled the lands of the Fertile Crescent, from the Nile River to western Iran from the imperial heartland along the Tigris River. Originating at the city of Ashur in the middle of the second millennium, the

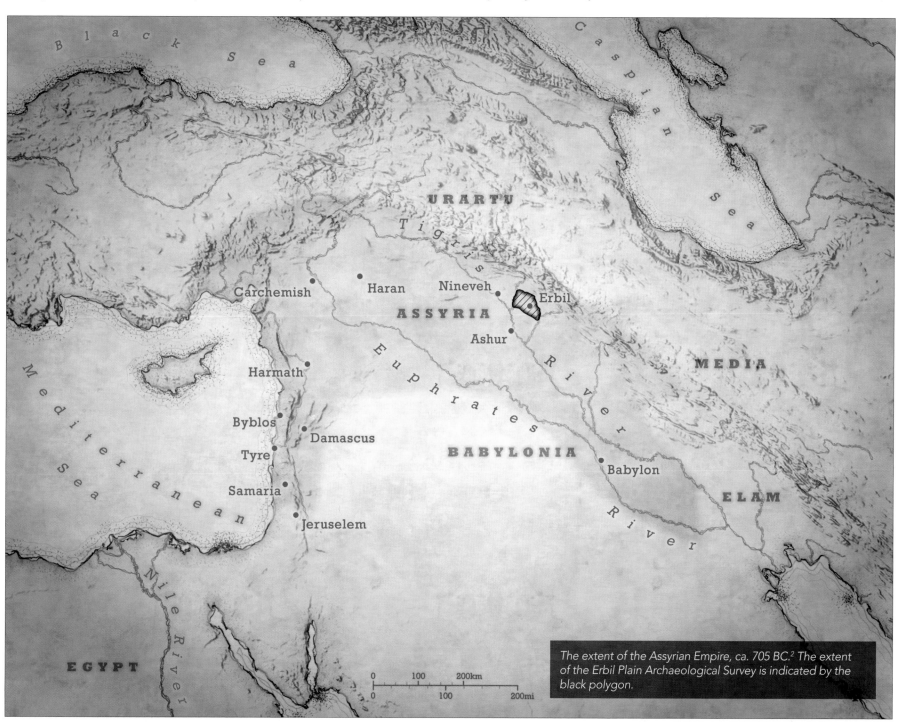

The extent of the Assyrian Empire, ca. 705 BC.[2] The extent of the Erbil Plain Archaeological Survey is indicated by the black polygon.

kingdom's fortunes grew and fell (but mostly grew) to include most of modern Iraq and Syria and large parts of Iran, Turkey, Jordan, Lebanon, and Israel. From Ashur, its kings relocated its capital to Nimrud, Khorsabad, and ultimately to Nineveh, which at the time of its destruction was the largest city in the world. These cities were supported by massive landscape projects—dammed rivers, long canals, and carefully engineered aqueducts. Most of these projects were covered by ideologically charged inscriptions and reliefs that made it clear to all the power of the Assyrian king and his divine supporters. But who inhabited these cities and landscapes?

In cuneiform inscriptions on the walls of their palaces, the kings themselves describe their conquests and how they removed entire kingdoms—their people, animals, and possessions—and brought them back to Assyria or other parts of the empire. Such inscriptions are inherently propagandistic and not to be read without skepticism, but in at least one case, the accounts of the victims of Assyrian deportation practices tell us the same story.

The heartland of ancient Assyria thus presents a series of interesting geographic questions. In what ways, and to what degree, did Assyrian kings and planners deliberately alter the landscape? Did they really deport several million conquered persons, and if so, where did they go? What impact did these forced migrations have on the landscape? Did these precocious transformations survive the political power behind them? As powerful as they were, the Assyrians were neither the first nor the last great power to dominate the Near East; how do their impacts compare with what came before them, and how did they structure those that followed?

Archaeologists' ability to answer these and other geographic questions about past civilizations have been revolutionized by the use of GIS. This chapter describes how a Harvard-led international team of archaeologists uses georeferenced drone imagery, historical satellite imagery, and good old-fashioned boots on the ground to investigate the traces of this once-mighty empire in its former core provinces, today the autonomous Kurdistan region of Iraq.

The Assyrian Empire was ruled from a sequence of capital cities on the Tigris River. King Ashurnasirpal founded the city of Kalkhu (modern Nimrud)[3] at the start of his reign in the ninth century BCE. A citadel mound held his palace and important temples. It was a vast lower town enclosed within a city wall (blue). Processional boulevards 20 m wide (red) have been found using satellite imagery.

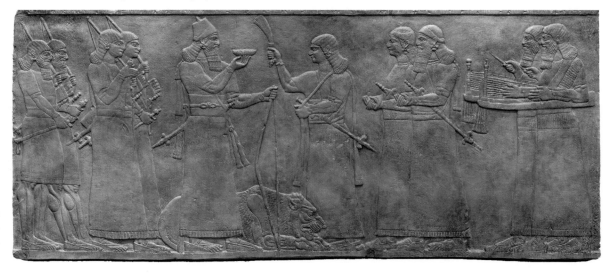

The Assyrian kings built great cities with large palaces, most of which were adorned with marble reliefs showing the king performing various royal duties. On this stone carving,[4] King Ashurnasirpal conducts a ritual after a successful lion hunt, with attendants and courtiers looking on.

LANDSCAPE ARCHAEOLOGY

The discipline of archaeology reanimates the past using its material remains. In public perception, these remains are artifacts—most often the detritus of daily life such as broken pottery, discarded stone tools, the waste of meals and mundane activities—but occasionally sensational finds such as the contents of royal tombs, palaces, and temples. Archaeologists find these objects via excavation and record them in 3D space. In more than 150 years of archaeology as an academic discipline, we have developed robust methods for turning them from data into interpretations.

In the later twentieth century, however, archaeologists realized that we cannot dig ourselves into answers for many important questions—particularly geographical questions about matters such as the migrations of hominids, the environmental impacts of agriculture, the origins of cities, and the nature of imperial landscapes. No excavation trench will ever be large enough.

For these and many other questions, the discipline of landscape archaeology has emerged. It shifts focus away from objects to a broader dataset of artifacts, sites, and landscape features—often captured from remote sensing platforms such as satellites, aircraft, and most recently drones—positioned and analyzed in time and 3D space. The origins of landscape archaeology were in the realization that the distribution of surface artifacts on a site was meaningful, and the arrangement of sites across a landscape could give important clues to the nature of past society, including its economic, political, and even religious structures. More recently, the spaces between sites (the traditional subject of archaeological study) have been recognized to contain a range of "off-site" features, such as tracks, roads, irrigation canals, field systems, and various symbolic monuments. These features are often more ephemeral than the sites that result from habitation, but they can be detected and mapped sometimes quite readily through the study of imagery.

Declassified intelligence satellite imagery from the 1960s reveals spectacular landscapes that are often invisible on the ground or since destroyed. The lines of light spots in this CORONA photograph[5] from February 1967 mark collapsed shafts leading down to subterranean water channels. These features fed water to the medieval town at lower left. Today they have been completely obliterated by industrial agriculture.

0   500 m

GPS device and CORONA photographs.

In the distance, a large mounded archaeological site rises over the Erbil Plain. In the foreground, a small low site is capped by a solitary abandoned structure. Today most of the plain is agricultural land, but it is rapidly developing as the city of Erbil expands.[6]

The tools of landscape archaeology have evolved radically in the past decades. In the twentieth century, archaeologists surveyed the landscape with a compass, pencil, paper, and their feet on the ground. Lucky ones might have access to aerial photographs. Landscape-scale phenomena are well studied from a vertical or remote perspective, and the availability of commercial high-resolution satellite imagery and declassified historical satellite imagery has been transformative. Around 2000, compasses began to be replaced by global positioning system (GPS) receivers; pencil and paper gave way to GIS software—first in the lab and later on mobile devices in the field.

Today, landscape archaeological research can be entirely digital, with precision global navigation satellite system (GNSS) receivers, smartphone-based mapping apps, imagery courtesy of a drone buzzing overhead in real time, and spatial data shared instantly with team members and even the public via cloud storage and web maps. Of course, we still do a great deal of walking; feet on the ground have yet to become obsolete.

Bashtapa from an altitude of about 400 ft.

Drones have revolutionized landscape archaeology, especially our ability to visualize ancient sites such as Bashtapa, as shown in the accompanying image. Here, team members Khalil Barzinji (left) and Nader Babakr oversee a drone mapping flight.

In rapidly developing areas, the traces of ancient land use are often threatened by the modern landscape. In this river valley, a 20-m-wide Assyrian dam was revealed by industrial gravel mining, which now threatens to destroy it.

THE KURDISTAN REGION OF IRAQ

Our research on the Mesopotamian past takes place in the modern Kurdistan region of Iraq. The Kurdistan region is predominantly mountainous, the western flanks of the Zagros Mountain chain that has been the traditional home of the Kurds and their ancestors for millennia. Farmers grew crops in narrow highland valleys and plains below the foothills, and pastoralists led their animals from high summer pastures to winter grazing on the plains near the Tigris River. Some towns existed, such as Rowanduz and Erbil, but the Kurds were predominantly a rural people.

At the start of the twentieth century, the Kurds lived as part of the Ottoman Empire, where they were restive but largely autonomous. After World War I, their lands were divided among the new countries of Turkey, Syria, and Iraq, also with communities in western Iran. Within the Republic of Iraq, the twentieth century was a time of fluctuating levels of strife between the Kurds in the north and the Arab-dominated national governments in Baghdad. It culminated in the genocidal Anfal campaign of the late 1980s, in which towns were gassed, approximately 3,000 villages were destroyed, and more than 50,000 men and boys were taken away, never to be seen again.

Autonomous Kurdistan region

Republic of Iraq

The Republic of Iraq and its autonomous Kurdistan region.[7]

As of this writing, the Kurdistan region of Iraq has rebounded. The rural countryside is being repopulated, and the cities of the Kurdistan region are being developed rapidly, particularly Suleimaniya and Erbil, the capital. The region is politically stable, safe, and welcoming to foreigners, including Western researchers. The Erbil Plain Archaeological Survey (EPAS) is a Harvard-led collaboration between foreign and local archaeologists in Erbil (population 850,000) in the Kurdistan region. It is identifying and mapping all archaeological sites on the plain of Erbil using all available tools and resources.

The renaissance in the Kurdistan region includes a desire to understand its past. In the twentieth century, the national governments in Baghdad promoted a national history, which focused almost exclusively on the civilizations of the southern plain between Baghdad and Basra. The Kurdistan region was left out of this history; few excavations were conducted, and foreigners were explicitly kept out of the region.

Today, the Kurds are interested in developing their own history, and the place to start is with documentation of the full cultural landscape of the region. Local archaeologists have been trained in traditional excavation methods, which are poorly suited for such a broad geographic undertaking. For this reason, they have been willing to partner with foreign research projects to develop a region-wide inventory of archaeological sites and historical places and to learn to use modern technologies such as drones, accessible historical imagery, and related geospatial tools and techniques.

At the center of the modern city of Erbil sits the qala (citadel) mound.[8] After six millennia of settlement, it is still crowned by traditional mud-brick houses. In 2016, it was named as a World Heritage Site.

A Royal Air Force plane took this oblique view[9] of Erbil in November 1938. It serves as an invaluable snapshot on which to build an understanding of the ancient landscape.

Traditional Kurdish lifeways revolved around agriculture and animal husbandry, which still characterize the rural plains around Erbil.[10]

SPYING ON ANTIQUITY

Archaeologists often fantasize about time machines that would allow us to witness firsthand the people and places of the past. Landscape archaeologists increasingly rely on the next best thing: historical remote sensing sources. The late twentieth century was a time of global development, particularly in the countries of the Middle East. As a result, the remains of past settlements and landscapes have often been badly degraded by the time archaeologists attempt to study them: cities and roads grow over them, agricultural developments plow them up, or hydroelectric dams flood them. If we can find aerial or satellite imagery that predates these developments, we can often reconstruct elements of the past that barely survive or no longer survive at all. Historical imagery has proven to be extremely powerful for this work.

Site identification depends heavily on historical remote sensing sources, especially U2 aerial photographs (1958–1960), KH-4 CORONA (1966–1968), and KH-9 HEXAGON (1972–1984) satellite photographs.

U2 aerial photographs (code named CHESS).[11] In the late 1950s, the United States flew missions over Middle Eastern countries from a base in Adana, Turkey. The photos were taken from an altitude of over 21,000 m and included vertical and near obliques and occasionally obliques up to the horizon. At best, they are better than 50 cm spatial resolution. The cameras ran continuously and covered a swath that was roughly 35 km wide along the plane's flight path. Eleven missions comprising over 50,000 frames have been declassified and their film deposited in the US National Archives. U2 images have been instrumental in identifying ancient sites and canal systems in the EPAS region.

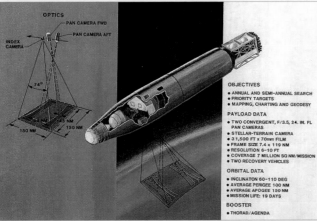

KH-4 satellite photographs (code named CORONA).[12] The first US intelligence satellite program acquired imagery from 1960 to 1972. Spatial resolution was much coarser than U2 (1–2 m in the later systems) but geographic coverage was much greater. CORONA was declassified in 1995, and its imagery has been available since 1998. Many scenes can be downloaded freely from the US Geological Society (USGS) EarthExplorer portal. It has emerged as a critical resource for Middle Eastern archaeology, serving as the basis for dozens of projects. EPAS relies heavily on Mission 1039 (February 28, 1967) to identify sites, and we use it as a base image for mobile GIS.

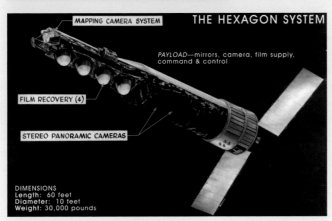

KH-9 satellite photographs (code named HEXAGON).[13] Running from 1971 to 1984, HEXAGON was the successor to CORONA, with a similar footprint but much higher resolution (ca. 50 cm) and more frequent target revisits. HEXAGON satellites flew over the Erbil Plain at least a dozen times. Like the U2 films, HEXAGON film is held by the US National Archives.

Because of the extreme geometric challenges of CORONA and HEXAGON, most scenes were georeferenced with a global third-order polynomial transformation. Scenes for our primary base imagery dataset, CORONA mission 1039 (February 28, 1967), were transformed using the spline method, which emphasizes local accuracy of specific ground control positions. Geo-corrected images are stored in a raster mosaic dataset in an ArcGIS geodatabase.

These source datasets are all panchromatic film—scanned into high-resolution grayscale raster imagery—which when incorporated into the geodatabase, become extraordinarily powerful tools for identifying ancient places. When a mud-brick settlement is abandoned, natural agencies will cause it to decay back into an earthen mound, with a lighter, looser soil texture than the natural soils that surround it. These anthropogenic soils often stand out strikingly as light areas on panchromatic photographs because they are drier and often host less vegetation. Conversely, depressed features such as canals and trackways collect moisture, and therefore promote vegetation growth; they appear as dark lines.

EPAS team members have georeferenced dozens of declassified historical images by using ArcGIS® Online basemap imagery as ground control. These visual signatures help researchers recognize potential ancient features, which are vectorized into points, lines, and polygons within the desktop software ArcGIS® Pro. In this manner, EPAS recognized over 1,600 potential sites, nearly 170 km of ancient canals, and over 340 km of premodern trackways before setting foot in the field.

But while remote sensing is clearly a powerful tool, it is still only the first step—these features must be confirmed on the ground. While rooted in old traditions of archaeology, the fieldwork described next has in its own way been revolutionized by geospatial technologies, most notably the use of highly accurate GPS receivers and other internet-connected devices that allow real-time access to maps and features in the geodatabase.

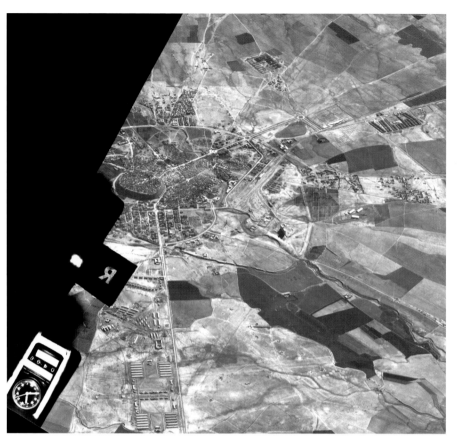

A U2 spy plane took this oblique view of Erbil in January 1960[14] as it banked sharply to the left. Nearly all the agricultural land visible in this photograph is now under the modern city. The analog clock face on the corner was a predigital means of what we would today call an image time stamp.

The Assyrian provincial capital city of Kilizu (modern Qasr Shemamok) shown in successive images taken in 1960, 1967, 1972, and 2018. The oval city wall is clear, with a citadel mound at its northern edge. The village of Saadawa, with a dozen houses, sat just to the south. Today, the site is being investigated by a French archaeological team, and people have returned to Saadawa village, which was destroyed by the Iraqi Army in 1987.[15]

FIELDWORK: MAPPING SITES ON THE GROUND

A durable pair of hiking shoes and a good hat are critical for the next step: field survey. Potential sites identified from aerial photographs and satellite imagery must be located on the ground, mapped, and sampled for chronologically sensitive artifacts.

Our field lab is the plain around the city of Erbil, capital of the autonomous Kurdistan region of Iraq. The modern city sprawls around an ancient citadel mound that encases within it the stratified remains of settlements going back an estimated 6,000 years. In 2011, Harvard archaeologist Jason Ur signed a contract with the Directorate of Antiquities to survey 3,200 sq km around the city—an area that was at one time the heart of the Assyrian Empire, and today is rapidly developing as Erbil becomes a successful modern city.

Preliminary identification of potential archaeological sites and landscape features was done by visual interpretation of remote sensing datasets, mostly panchromatic aerial and satellite photographs from the late 1950s through 1980s. Because eroded mud brick has a loose soil structure that sheds moisture more quickly than the natural soils, potential sites appear as light discolorations on a darker background. These sites were vectorized as polygon features. Landscape features such as premodern tracks and irrigation canals are often slightly depressed and therefore appear on imagery as dark lines. These features were vectorized as polylines. All datasets were uploaded as hosted feature layers in ArcGIS Online, where they could be shared with team members.

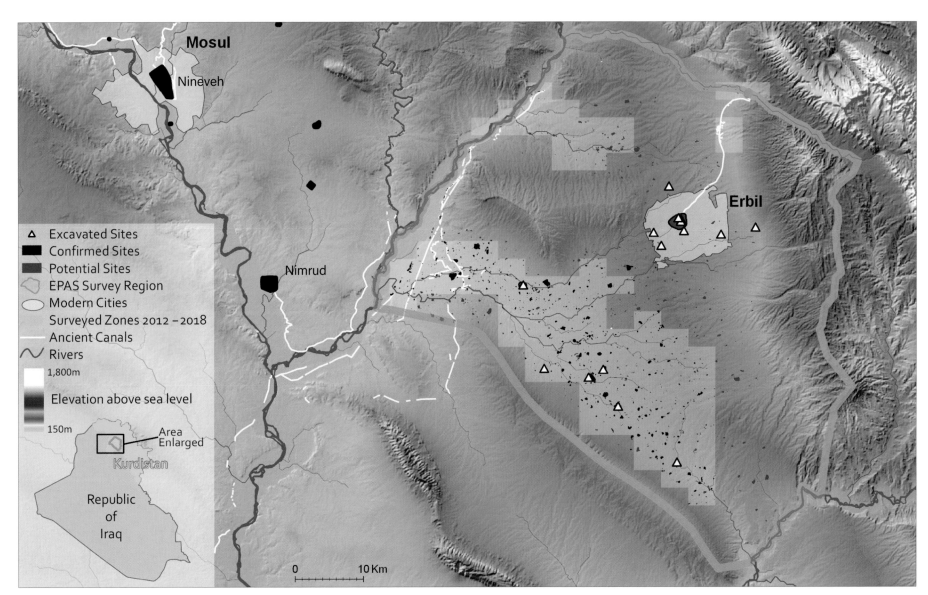

EPAS is examining 3,200 sq km of historically rich landscape around the city of Erbil, formerly the core of the Assyrian Empire.[16] Sites visited and confirmed are shown in black; purple sites are potential ancient places that have not yet been visited. Since 2012, the survey team has covered some 930 sq km of our 3,200 sq km project region. In the early first millennium BC, the Erbil Plain was part of the core of the Assyrian Empire, which had capital cities at Nimrud and Nineveh.

The survey region is divided into blocks of 10 sq km. Intensive archaeological surveys often opt to walk transects across the landscape, often at close intervals. Unfortunately, the growth of Erbil does not allow us this luxury—the city and other developments are happening too fast. We visit our sites in vehicles, which enable us to cover large areas, albeit at a low intensity. We have found, however, that nearly all sites are visible in one or more of our imagery sources.

When the team reaches a site, its members spread out across it to collect chronologically sensitive artifacts from the surface. These are overwhelmingly broken pieces of pottery, the detritus of daily life until the arrival of inexpensive plastics in the twentieth century. If the site is large, its surface will be divided into two or more areas, ideally based on topography. Increasingly, we rely on our drone program to produce digital terrain data for this purpose. The spatial distribution of artifacts from different time periods is the basis for our estimations of site size at various points in the past.

While some team members collect artifacts, others fan across the site to map its boundaries and the edges of the collection areas and to take elevation points of the site and its surrounding plain. When cellular data allows it, each team member's additions to our geodatabase are immediately visible to the rest of the team, wherever they are on site. We attempt to describe and document the condition of the site, especially any damaged or looted places, and to document any impending threats that it might face in the future. Early in our project, this documentation involved pencils, notebooks, and ground photography; now it features centimeter-resolution orthophotos, low-level drone oblique photographs, and voice-dictated digital notes. When the team has finished at any given site, they have often collected dozens of waypoint observations, dozens of artifacts, and hundreds of unmanned aerial vehicle (UAV) photographs. The team is then ready to move on to the next site.

Potential sites recognized in historical remote sensing sources must be confirmed on the ground. Sites and other datasets are organized into web maps via ArcGIS Online and visible to all team members through the Explorer for ArcGIS® or Collector for ArcGIS® apps, depending on the team member's ability to edit or only view spatial data. Most team members use Bluetooth-enabled Garmin GLO GPS receivers; others who need to make precise measurements (e.g., ground control for drone photogrammetry) use Trimble R1 units with real-time correction via the ViewPoint RTX service. Team members switch between Esri imagery (various dates, but mostly the last five years) and CORONA satellite photographs (1967) as base imagery.

Team members collect surface artifacts within discrete subareas of sites, using the GIS apps on their mobile devices to guide them, and they mark the boundaries of the site based on the declining density of artifacts.

Drone-derived 3D models now guide the team's collection strategies.[17] A digital elevation model (DEM) allowed us to subdivide the surface of this small mound in advance, so that team members know precisely where to collect artifacts. Surveyors wear GPS watches so their movements (coded by color) are mapped.

Surface artifacts are cleaned and dried in the sun before they are analyzed.[18] Pottery can be chronologically sensitive, allowing us to determine when people lived on a site.

DRONES IN ARCHAEOLOGY

For nearly all its history, the villages, towns, and cities of ancient Assyria were constructed of earth. In this stone- and tree-poor region, sun-dried mud bricks were the best and only material available for peasant and king alike. Mud-brick architecture will not last for more than a few decades, and eventually, any structure must be leveled and a new one rebuilt in its place. As a result, settlements grow vertically when they remain occupied for centuries or millennia. When they were abandoned, they decayed back into earthen mounds. Their volume can tell archaeologists something about their duration and continuity. Furthermore, their topography can give clues to settlement histories. In the past, archaeologists created topographic site plans using standard surveying total stations, a slow process that involved two persons, a prism pole, a bubble level, and a great deal of patience.

Drones have changed all of this. In the past few years, they have become inexpensive and easy to use, and flight planning and photogrammetric software has followed suit. On the Erbil Plain, most sites can be planned and flown in ten or fifteen minutes. Our drones fly at a height that produces orthophotos with a resolution of 1 to 2 cm and digital terrain around 5 cm.

The project first adopted drones in 2016, and they quickly became much more than a novel source of interesting views—they are a standard part of basic site recording. Our drone protocol unfolds over three days. On the first day, the drone team visits a site, lays out and records ground control positions, and flies a preplanned mission. On the second day, the images and ground control points are brought together in photogrammetric modeling software to produce orthophotos and terrain models, which are then used to plan the team's surface collection areas. These polygons are added to the hosted feature layer in ArcGIS Online. On the third day, the collection team revisits the site to make precise artifact collections, guided by the collection area polygons on their GPS-enabled mobile devices. The team repeats this process on most sites.

Usually, increasing the speed of research requires a corresponding reduction in accuracy. This combination of drone survey and cloud-based mobile GIS allows the teams to move more quickly but with enhanced spatial precision. This is a critical combination, since we are often working in advance of new development projects that might damage or destroy the archaeological landscape before our next field season.

In 2018, the survey added a fixed-wing UAV. It has a flight time of around 45 minutes and a precision GPS on board. Here a SenseFly eBee Plus sits on the ground near an Erbil Plain mound.

Team member Dr. Bjoern Menze (Technische Universität München) devised our original drone mapping protocol in 2016.[19]

Team member Pshtiwan Ahmed sends up the project's quadcopter to map an archaeological site.

This site at Qatawi shows the encroachment of the surrounding modern town. Rapid development in the region lends urgency to the archaeological mission.

Team member Shilan Ramadan records the position of a nylon sack in the field. High-contrast debris can often be used as ground control points in situations where the team must move too quickly for formal photogrammetry targets.

The details of this archaeological site, which include military trenches, a modern cemetery, and encroaching housing, would have taken a week or more to map using traditional methods. Our drone program mapped this site in 15 minutes with 300 photographs.

DATA COLLECTION AND PROCESSING

To reconstruct its settlement history we must subdivide any particular mound into artifact collection areas. Girdi Abdulaziz is a sprawling complex of high mounds and low rises near Erbil. It lies outside the Erbil core urban area where overdevelopment is taking place. This site will serve as an example of exploring the advantages of drone data.

For large and complex sites like this one, topographic mapping is done days before collection via drone photogrammetry. The drone team has used quadcopters since 2016, and in 2018 the project added a fixed-wing drone for greater areal coverage. Vertical coverage is planned via drone apps with 60–70 percent overlap and generally at a height of 75–100 m, depending on the area to be covered and available battery life. For a detailed list of what hardware was flown, see the endnotes.[20]

For reasons of shadows and sunlight intensity, flights are made between 7 a.m. and 10 a.m. local time. Image GNSS positions from eBee flights are post-processed with reference to Continuously Operating Reference Station (CORS) data from a base station in south Erbil to around 8–10 cm accuracy, but quadcopter flights require manually collected ground control points via Collector and a Trimble R1 unit. Team members place sheets of paper within the flight area, or opportunistically identify high-contrast ground objects (most often rubbish such as black or white plastic bags) on the order of 15–20 cm.

The same afternoon and evening, drone images from the day's missions are processed in Drone2Map® for ArcGIS, most often as a batch that runs overnight. Output orthophotos have a spatial resolution of 1–2 cm, and digital surface models (DSMs) have around 4–8 cm. The next day, the DSM is used to plan subareas for artifact collection on complex sites. The subarea polygons are created in ArcGIS Online–hosted feature layers and are immediately available to all team members to guide field navigation and collection.

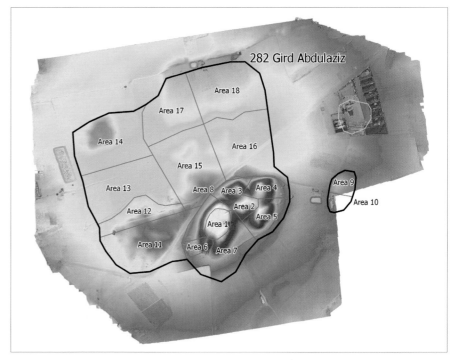

Girdi Abdulaziz 15 cm resolution DEM with labeled collection areas.[21]

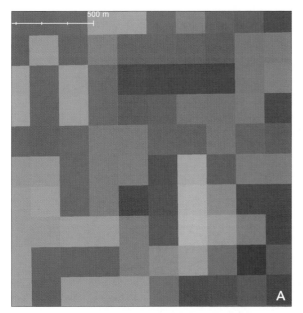

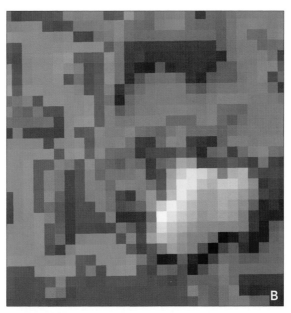

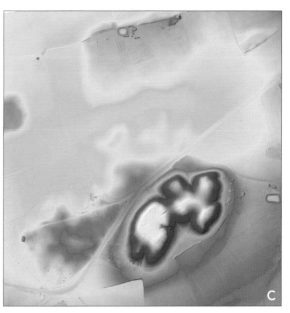

Available terrain datasets such as the 90 m (A) and 30 m (B) Shuttle Radar Topography Mission [SRTM] terrain models are too coarse to guide us effectively. SRTM is data that originated from a specially modified radar system that flew on board the space shuttle Endeavour *in 2000, and that produced global digital elevation models at a 30 m resolution. In contrast, a UAV flight collected 1,236 overlapping vertical and oblique photographs that were used to construct the 15 cm digital terrain model (C). All three terrain models cover the same extent.*

PROCESSING DRONE DATA

The use of drones has revolutionized the work of landscape archaeologists by providing a means of putting a camera in the air at relatively low altitudes. This technlogical leap forward delivers imagery that is higher resolution by orders of magnitude than satellite imagery. And while the imagery itself is inherently useful, its power is multiplied when processed through Drone2Map for ArcGIS, which streamlines the creation of professional imagery products from drone-captured still imagery for visualization and analysis in ArcGIS.

Through the magic of software, a series of individual photos taken from the drone are overlapped and stiched together to create seamless imagery products, including orthomosiacs; 3D models (aka DEMs); and point clouds, which enable analysis of natural and built-up features, including volumetric measurements, change detection, lines of sight, and obstructions.

Drone missions like this one at Girdi Abdulaziz include autonomous transects planned for vertical overlapping coverage and also manual flights for oblique views.[22]

0   50

An orthomosaic image is an aerial photograph that has been "orthorectified" (geometrically corrected) such that the scale is uniform and the image is placed in geographic space, which means it can be loaded into a GIS and combined and analyzed with other geodata, including vector. As such, an orthomosiac can be used to measure true distances, because it is an accurate representation of Earth's surface, having been adjusted for topographic relief, lens distortion, and camera angle.

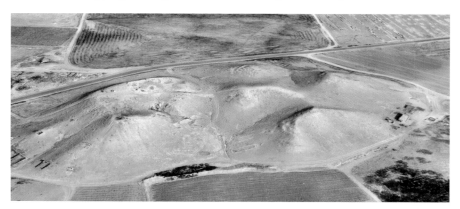

This oblique view was created by draping a geographically correct orthophoto over the 15 cm digital elevation model of Girdi Abdulaziz.

A revisit of the site in January 2019 revealed not the typical dry, brown views, but a much greener landscape than on previous expeditions.

CASE STUDY: DISCOVERING A LOST KINGDOM

After two centuries of archaeology, there are still "lost cities" awaiting rediscovery. Modern archaeology increasingly makes such discoveries from space.

In the 1970s, the Iraqi State Board of Antiquities and Monuments recognized a small mound of the Middle Bronze Age (ca. 2000–1500 BC) at a place called Kurd Qaburstan ("cemetery of the Kurds"), a short drive south of Erbil. In preparation for the first season of field research, project director Jason Ur identified this and many other sites on a CORONA satellite photograph from 1967. What made Kurd Qaburstan stand out was not its ruin mound but rather the odd linear feature that surrounded it. It seemed far too straight to be a natural watercourse, and in

some places, it was paralleled by areas of light discoloration. It had the potential appearance of a massive wall enclosing the site. At 100 hectares, it would therefore be one of the largest cities of its time.

It happened that archaeologists and ancient historians had been searching for just such a place for decades. It was known from other ancient sources that two kings had temporarily allied themselves to attack the kingdom of a third king, whose capital, known as Qabrā, was located somewhere on or near the Erbil Plain. One of the two conquerors had even commissioned a depiction of Qabrā and its city wall on a stone monument, as shown in the drawing.

The American CORONA spy satellite flew over Kurd Qaburstan on February 28, 1967. The light discolorations are areas of mounding, which contain collapsed mud brick. The yellow line highlights the curving oval shape of the buried city wall [23].

0   500 m

Kurd Qaburstan has a mound at center, today covered by a cemetery. These fields cover an area of dense urban housing that flourished almost 4,000 years ago.

A farmer near Baghdad found this stone monument in a field. This drawing [24] of Qabrā shows King Dadusha of Eshnunna defeating the king of the walled city.

The city of Qabrā had been lost to history, but now just such a walled place was visible in a declassified spy satellite photograph at Kurd Qaburstan. The survey team arrived at the site in July 2012 with a sense of excitement. The hypothesized city wall was a low rise in the midst of agricultural fields in some places, but in other spots, it was invisible. The surface yielded broken pottery from many different time periods, but most frequently occurring were artifacts of the Middle Bronze Age—and scattered across the whole of the site.

The dating of this city seemed assured, but we needed more confirmation of its wall. That confirmation has come in two forms. One confirmation was topographic. Using more than 1,000 drone photographs, we created a 3-cm resolution terrain model of the site, in which the wall could be traced around the entire site and even under a nearby village. Further confirmation came from colleagues at Johns Hopkins University, who began excavations at the site in 2013. Their project included a magnetometry survey, which revealed stunning images of the buried wall, including rectangular towers at 20-m intervals, as well as a dense network of streets in urban neighborhoods beneath the fields. Excavations confirmed that the wall had been constructed in the Middle Bronze Age.

We cannot be 100 percent certain that this place is indeed Qabrā until some ancient inscription from the site confirms it; that will have to await future results of the excavation.

In raking morning light, the mounds at the center of Kurd Qaburstan stand out.[25]

The magnetometry survey was carried out by Dr. Andrew Creekmore, who found a dense urban grid and a city wall with towers at 20-m intervals along the city's northern edge.

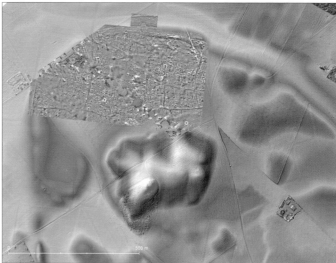

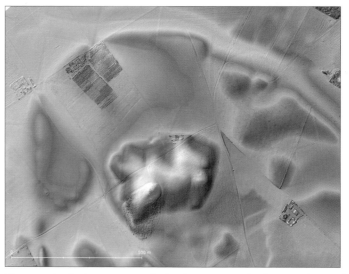

The geophysical survey revealed a dense network of streets beneath the fields and a thick city wall with towers. The massive bulk of the ancient city at Kurd Qaburstan is visible from this drone-derived DEM.

REVEALING EVIDENCE OF FORCED MIGRATION

One focus of the survey is the investigation of possible forced immigration under the kings of the Assyrian Empire of the first millennium BC, but forced migration did not end with the collapse of the empire; it has continued as a tool for control over rebellious populations globally up to the present. Village destruction and population relocation was carried out by the Iraqi Army under the regime of Saddam Hussein's Ba'ath Party up to its fall in 2003. Victims included various minority groups that were determined to be disloyal, including the Marsh Arabs of southern Iraq and the Kurds in the mountains of the north.

In the late 1980s, the Iraqi Army moved to punish the rural populations of Kurdistan, who stood accused of supporting Iran during the Iran–Iraq conflict. This genocidal campaign is most notorious for the use of chemical weapons against civilian Kurds in 1988, but it began the previous summer, with the depopulation and destruction of hundreds of rural villages across Kurdistan, including nearly every village of the Erbil Plain.

As we locate and map sites of much greater antiquity, our team members often encounter the ruined villages of the 1980s. As peace has returned to Kurdistan, many of them returned to life. Although our research objectives are deeper in the past, we feel obligated to give these villages the same level of attention that we apply to more ancient places in the hopes that they will not be forgotten.

Today the remains of Biryam Malak appear as a series of low mounds on the plain.[27]

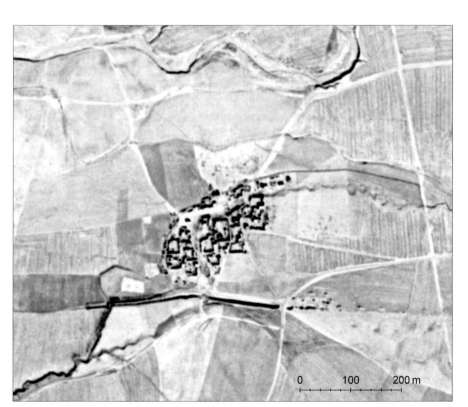

In February 1972, Biryam Malak was a village of at least a dozen courtyard houses, as shown by this HEXAGON satellite photograph.[26]

In April 1987, the village was evacuated and destroyed by the Iraqi Army, along with nearly all other villages on the Erbil Plain. This orthophoto from September 2017 shows the remains of the bulldozed houses.

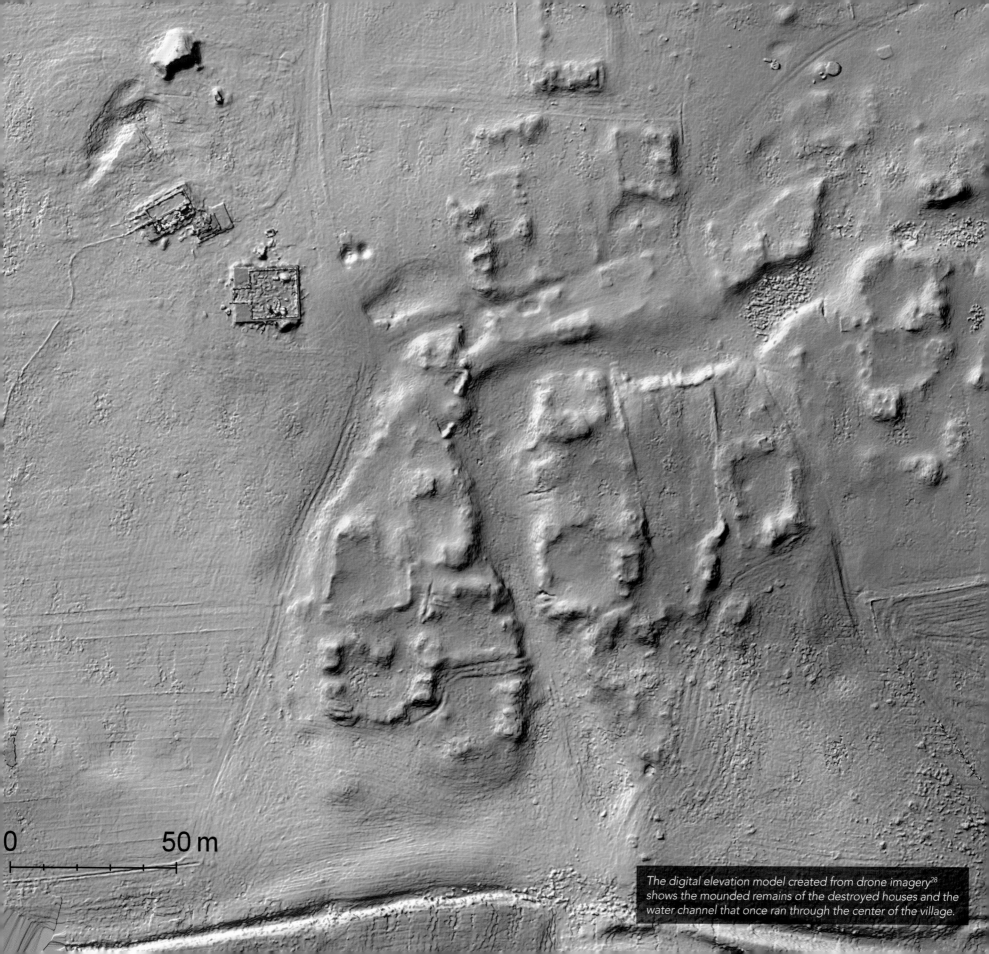

0 50 m

The digital elevation model created from drone imagery[28] shows the mounded remains of the destroyed houses and the water channel that once ran through the center of the village.

METHODS AND RESULTS

Over the course of five field research seasons, the Erbil Plain has revealed an amazingly rich archaeological landscape. Leaving aside for the moment the "big question" about the landscape impacts of formation of the Assyrian imperial core, we can consider some numbers:

- 930 sq km visited on the ground
- 585 sites identified and mapped (out of 841 potential sites visited, a success rate of 70 percent for our remote sensing site identification program)
- Almost 350 km of ancient trackways identified
- 140 premodern canal segments mapped, extending almost 170 km
- 1,739 bags of artifacts collected, and almost 55,000 artifacts analyzed
- Over 80,000 drone images acquired, encompassing 61 percent of our sites to date

Ultimately, we intend to describe the history of human settlement on the Erbil Plain, from the first sedentary farmers up to the modern era. To do so, we must spend much time analyzing the artifact collections, and at present, we have done so only in a preliminary manner. Nonetheless, a historical pattern is emerging. The earliest villages were small and few, and they adhered closely to permanent water sources. Starting in the fourth millennium BCE, site numbers and maximum site sizes began to grow as the social and political structures necessary for urbanization emerged. The first unambiguously urban place appeared around 2500 BCE.

Sometime after the rise of the Assyrian Empire around 900 BCE, the plain witnessed the greatest expansion of settlement in its history—both in terms of the number of sites and probably also in terms of population. The number of settled places spiked to a level not seen before, or since. With a few urban exceptions, these Assyrian settlements were very small—possibly the villages of forced migrants from elsewhere

in the empire, as depicted on the walls of the royal palaces. In the phase that followed the collapse of Assyrian power, settlement numbers receded dramatically. It is tempting to think that perhaps these deportees (or their descendants) chose to return home once the Assyrian army was no more.

The Assyrian settlement expansion occurred with some monumental transformations of the landscape. One artificial canal moved water almost 5 km across a watershed, ending in a 300-m-wide basin. Its appearance on CORONA satellite imagery is impressive but does not do it full justice; its 100-m width and 8-m depth are best appreciated through drone oblique imagery. It must have required the excavation of nearly 4.5 million cubic meters of earth at a time long before mechanization. North of Erbil, our drone program discovered the remains of a 20-m-wide dam across a river, which would have diverted its flow into a subterranean channel toward Erbil. Smaller irrigation features have also been recognized, and many more have probably been removed by development.

The imperial impact on the landscape is continuing to emerge. What has become clear is that archaeological research on the scale of empires needs geospatial tools such as these. Imagery sources, especially historical imagery, are especially powerful, both in the lab and in field navigation and site mapping. The best circumstance comes when historical imagery can be used in tandem with up-to-date commercial imagery and on-demand drone-derived orthophotos.

Our archaeological survey has far more work to do. We still must confirm and record almost 900 potential sites. Time is not on the side of the historical landscape, and Erbil continues to thrive and expand (see story map linked at GISforScience.com). Fortunately, the partnership between foreign and local archaeologists is strong, and our geospatial tools and methods are in place.

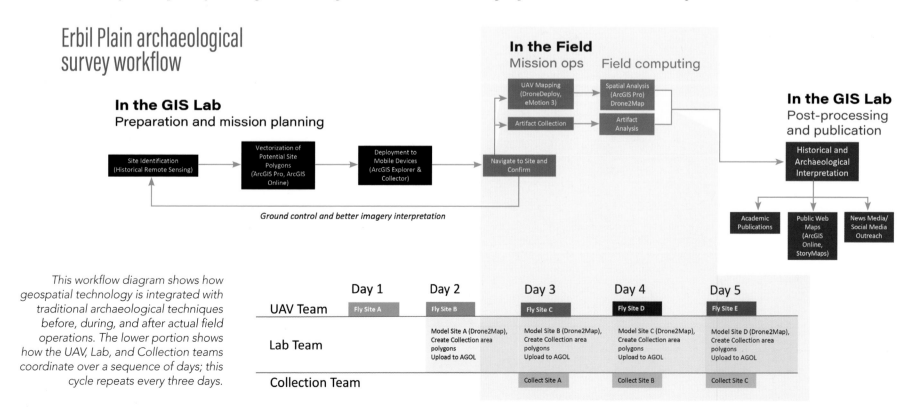

Erbil Plain archaeological survey workflow

This workflow diagram shows how geospatial technology is integrated with traditional archaeological techniques before, during, and after actual field operations. The lower portion shows how the UAV, Lab, and Collection teams coordinate over a sequence of days; this cycle repeats every three days.

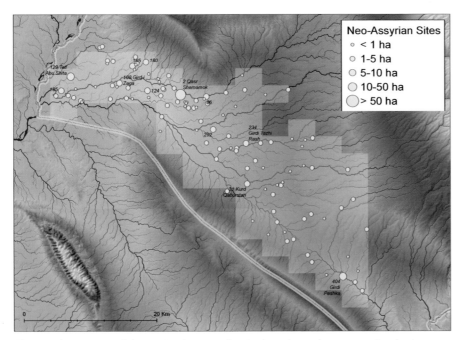

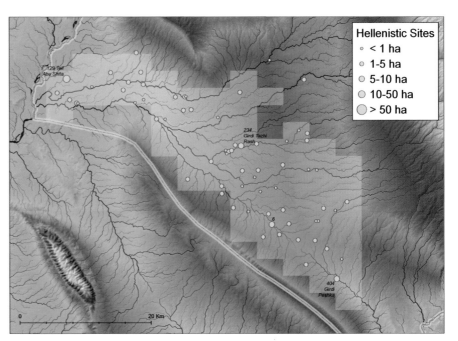

The southwest part of the project's research area has almost been completely surveyed. At the time of the empire, it hosted the densest scatter of villages and towns in the region's premodern history, including the city of Kilizu (modern Qasr Shemamok). Pink indicates the boundary of the survey research area; transparent yellow areas have been fully surveyed (to 2018).

After the conquests of Alexander, the plain looked radically different, with many settlements abandoned or reduced in size. The pattern of settlement did not survive the dissolution of imperial power.

Assyrian labor management transformed the hydrology of the Erbil Plain. This 100-m-wide, 8-m deep canal (shown here in a 1967 CORONA satellite photograph) would have required the excavation of 4.5 million cubic meters of soil by human labor. At its end was a 300-m-wide collection basin (inset, a 1972 HEXAGON satellite photograph).

THE ARCHAEOLOGISTS

The pursuit of the past in the Middle East is a long story, mostly narrated by archaeologists from great Western institutions. The earliest expeditions were sponsored by imperial governments via national museums. The collections of the British Museum and the Louvre, for example, derived from this tradition of exploration. In the twentieth century, the institutions were increasingly European and American universities; the finds remained in their countries of origin (now all independent of colonial control) but the tools, techniques, funding, and direction still originated in the West. In the twenty-first century, all of this is changing rapidly.

EPAS is led by Jason Ur, a professor of archaeology in Harvard's Department of Anthropology. Ur had led surveys of ancient landscapes in Syria, Iran, and Turkey. These projects had revealed to him landscapes that seemed to have emerged from the bottom up—the earliest cities that seemed to have self-organized around the interests of urban farmers and pastoral nomadic landscapes that took form from the cumulative decisions of small camping groups. The landscape of ancient Assyria, now in the Republic of Iraq, seemed to present a radically different case: an imperial core with hints of being planned from the top in all aspects. Iraq has been a dangerous place since the 2003 invasion and ousting of Saddam Hussein, but the autonomous Kurdistan region provided a secure and welcoming environment for research. It has proven to be an ideal place to test ideas about the geography of the Assyrian imperial core.

Our project started in 2012 with several co-directors, including Dr. Jessica Giraud, a French landscape archaeologist, and Dr. Lidewijde de Jong, a Dutch Classical archaeologist. Our official representative from the Kurdistan region's Directorate of Antiquities was Khalil Barzinji, an Erbil-based archaeologist with a degree in archaeology from the University of Baghdad. In this way, EPAS initially fit the twentieth-century model of foreign expeditions.

Starting in 2018, the survey project maintains all its spatial data in the cloud. Team members use ArcGIS Explorer and Collector for navigating and recording via smartphones and other handheld devices. With project data hosted in the cloud, all team members have immediate and total access to our spatial data.

Since that first year, the team's composition has evolved. The modern city of Erbil has grown outward, and, in the process, it is covering or threatening the historical landscape. It is not possible for a small team of Western researchers visiting for a couple of months to document its past before it disappears under new roads and buildings. We have, therefore, trained Khalil and other young local archaeologists in the techniques of landscape archaeology, so that they can participate as equal team members and conduct this work in the absence of the foreign team. Advances in mobile GIS and cloud-based data sharing have been instrumental—our database is now accessible via smartphone apps to every team member. Khalil directs our drone documentation program and pilots drones over sites for the Directorate of Antiquities when EPAS is not in the field. In 2018, our team was majority Kurdish and is now co-directed by Nader Babakr, the Director of Antiquities for Erbil Governorate. Our research is a true partnership between foreign and local archaeologists, what we hope will be a model for future work in this region.

Most field days start with a briefing by project directors Jason Ur and Nader Babakr (third and fourth from left, respectively.)

ENDNOTES

This research was funded by the US National Science Foundation (Award #1261118), the National Geographic Society, and the Dumbarton Oaks Research Library and Collection. We gratefully acknowledge the support of the Directorate of Antiquities of the Kurdistan Regional Government and the KRG's representation in the US.

1. Image courtesy of the Erbil Plain Archaeological Survey (EPAS).
2. Map by John Nelson, Esri.
3. Map by Jeffrey Blossom, Harvard University.
4. Image courtesy of Semitic Museum: (modern resin copy of original plaster cast courtesy of the British Museum).
5. CORONA image courtesy of USGS.
6. All images on this page courtesy of EPAS.
7. Map by Jeffrey Blossom, Harvard University.
8. Image courtesy of EPAS.
9. Courtesy of the Royal Air Force.
10. Image courtesy of EPAS.
11. US Air Force photo. Reproduced with the permission of Air Force.
12. Image of KH-4B Corona courtesy of National Reconnaissance Office.
13. U2 and HEXAGON photos courtesy of the US National Archives and Records Administration (NARA). CORONA photograph courtesy of USGS. Basemap imagery courtesy of Digital Globe.
14. Photo courtesy of the US National Archives and Records Administration (NARA).
15. Images courtesy of USGS.
16. Map by Jeffrey Blossom, Harvard University.
17. Map by Jeffrey Blossom, Harvard University.
18. Image courtesy of EPAS.
19. All images on this page courtesy of EPAS.
20. All images on this page courtesy of EPAS.
21. DJI quadcopters (Mavic Pro and Phantom 4) since 2016; in 2018 the project added a fixed-wing senseFly eBee Plus for greater areal coverage.
22. All maps on this page courtesy of EPAS.
23. CORONA image courtesy of USGS.
24. Drawing courtesy of Robert Rollinger, University of Innsbruck.
25. All images on this page courtesy of EPAS.
26. HEXAGON image courtesy of USGS.
27. Drone images on this page courtesy of EPAS.
28. All images on this page courtesy of EPAS.

From 1967 to the present, the city of Erbil has grown from a large town to the urban capital of the Kurdistan region. An interactive version of this slider comparison app can be accessed at GISforScience.com.

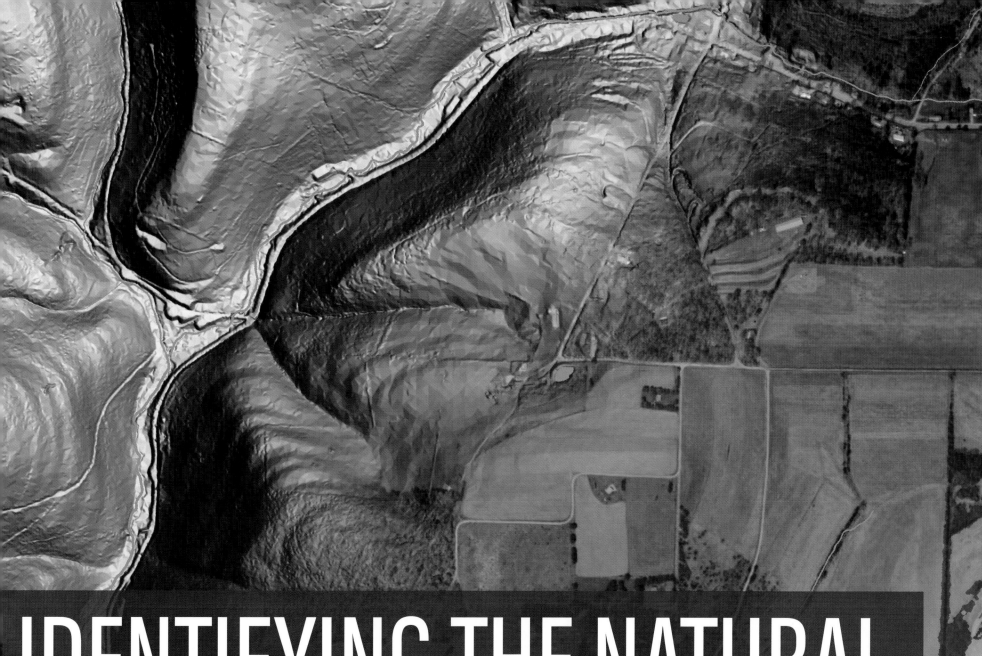

IDENTIFYING THE NATURAL EFFICIENT FRONTIER

To improve conservation efforts across the entire US, scientists are leveraging artificial intelligence and satellite imagery within GIS across large landscapes to find the very best places for restoration.

By Jeff Allenby, Chesapeake Conservancy; and Lucas Joppa and Nebojsa Jojic, Microsoft Research

This blended image of the Chesapeake Bay watershed is part of a massive high-resolution database that has allowed conservation agencies to model potential conservation opportunities.

THE CHESAPEAKE BAY WATERSHED: A COMPLEX ENVIRONMENTAL SYSTEM

The Chesapeake Bay is the largest estuary in the United States and the third largest in the world. Its 64,000-square-mile (165-square-kilometer) watershed is home to more than 18 million people and more than 3,600 species of plants and animals. Primarily because of land runoff issues, the health of the bay declined steadily during the twentieth century, resulting in the need for a focused and coordinated effort across the six watershed states and Washington, DC, to improve habitat and water quality.

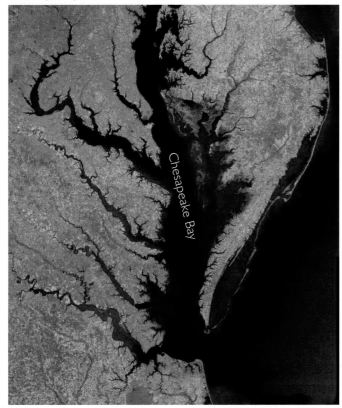

The Chesapeake Bay Program (CBP), a state-federal partnership led by the US Environmental Protection Agency (EPA), has guided restoration efforts in the Chesapeake during the last three decades, setting progessively more sophisticated restoration goals for each state based on a series of models representing the natural and physical processes of the watershed.

In 2010, the program partnership established a set of brand-new water quality goals—Chesapeake Bay total maximum daily load (TMDL)—a comprehensive "pollution diet" designed to restore clean water in the Chesapeake and the region's streams, creeks, and rivers by 2025. By dividing the Chesapeake watershed into thousands of subwatersheds, the team of state and federal scientists and researchers applied the TMDL thresholds to assess each subwatershed's water quality goals. Given the large scale and complexity of the problem, GIS plays a crucial role in the work.

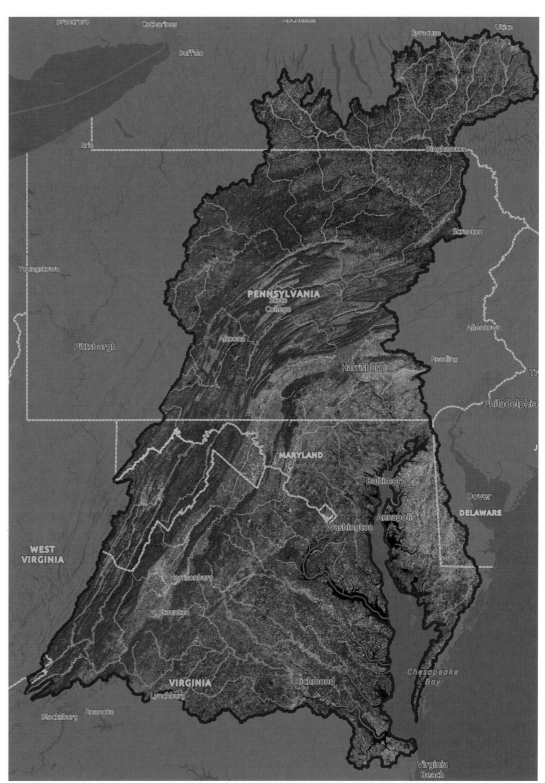

From a jurisdictional standpoint, managing pollution flows into the Chesapeake Bay watershed is a massive challenge. The watershed drains parts of six states—New York, Pennsylvania, Maryland, Virginia, West Virginia, and Delaware—plus Washington, DC. Its total area exceeds that of 27 US states.

A PARTNERSHIP APPROACH TO RESTORATION

In 2017—roughly halfway into the 15-year project—the CBP conducted a midpoint assessment to determine how much progress had been made and where additional efforts needed to be focused. This assessment found that the watershed states met the interim goals of achieving 60 percent of the necessary reductions, compared with 2009 levels, for sediment and phosphorus reductions, but did not meet the nitrogen reduction goal. In particular, the CBP found that Pennsylvania lagged in meeting its commitments to reduce loads from its urban and agricultural runoff, as well as suburban stormwater sources. Each state is revising its goals and strategies to meet the water quality standards of the TMDL by 2025.

To help meet the TMDL water-quality goals, each of the watershed states signed on to the Chesapeake Bay Watershed Agreement, which defines 10 goals that, if achieved by 2025, indicate a healthy and thriving Chesapeake Bay watershed. Each goal links to a set of management strategies and measurable outcomes with deadlines that are used to assess how each state is achieving success.

Partners throughout the bay's watershed have focused restoration efforts on implementing the 30 strategies outlined in the 2014 Chesapeake Bay Watershed Agreement. While each strategy is unique, they aim to create quantifiable metrics that allow CBP to track progress toward the goal of a healthy Chesapeake. Many of these metrics are simple to measure, but a number of strategies rely solely on quantifying the amount of effort that has gone into a goal, such as totaling the number of stream miles of riparian forest restoration that occurred in the last year.

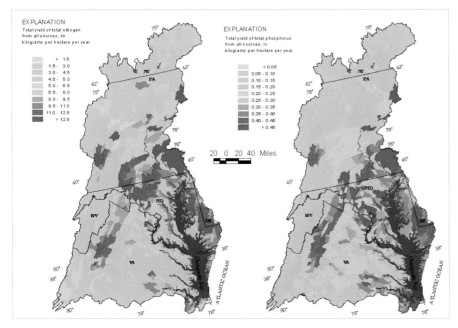

Existing modeling frameworks (like these early GIS maps from a 1992 assessment) have allowed the CBP to identify priority watersheds needing the highest pollution reductions, but local partners did not have the information to identify specific projects that would make the most progress toward these goals.

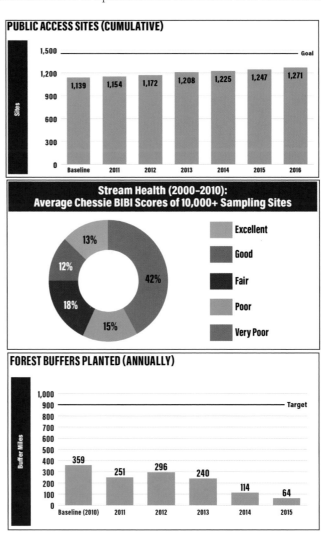

Tracking outcomes for each strategy varies and might detail cumulative metrics or the results of field assessments. In many cases, progress is calculated based on the amount of effort expended each year because appropriate data did not exist to measure implementation in a more comprehensive way (Chesapeake Conservancy 2017).

	Agriculture	Urban/Suburban	Wastewater	Trading/Offsets
Delaware	Enhanced Oversight	Ongoing Oversight	Ongoing Oversight	Ongoing Oversight
District of Columbia	Not Applicable	Ongoing Oversight	Ongoing Oversight	Ongoing Oversight
Maryland	Ongoing Oversight	Enhanced Oversight	Ongoing Oversight	Ongoing Oversight
New York	Ongoing Oversight	Ongoing Oversight	Enhanced Oversight	Ongoing Oversight
Pennsylvania	Backstop Action Levels	Backstop Action Levels	Ongoing Oversight	Enhanced Oversight
Virginia	Ongoing Oversight	Ongoing Oversight	Ongoing Oversight	Ongoing Oversight
West Virginia	Ongoing Oversight	Ongoing Oversight	Ongoing Oversight	Ongoing Oversight

The 2017 Midpoint Assessment found that most watershed states are in need of additional oversight to stay on track to meet their 2025 water-quality goals.

INVESTING ALONG THE EFFICIENT FRONTIER

Achieving an optimal balance between risk and reward—a priority across many fields and industries—also applies to environmental management. Frameworks and strategies from finance and investing, for example, are useful in complex environmental scenarios like the Chesapeake watershed problem.

In finance, investors have historically evaluated companies and the markets

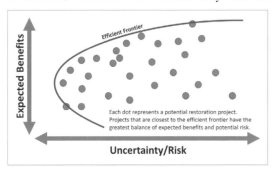

in which they operate to better understand the likely performance of a stock and gauge its likely risks and returns. In recent decades, because of advancements in computing and data availability, these evaluations increasingly use large datasets to quantify a market's behavior to gain deeper insight into a stock's expected performance.

Originally proposed by Harry Markowitz in 1952, the "efficient frontier" represents a hypothetical set of investments that offers the highest expected return for a

defined level of risk or the lowest risk for a given level of expected return. By keeping investments near the efficient frontier, investors can maximize their returns while managing the associated risks. Integrating the concept of the efficient frontier into environmental decision making can dramatically improve our ability to invest in the highest performing, most cost-effective projects.

For example, environmental modeling has historically focused on identifying the general landscapes where action is most needed to improve conditions. This approach has helped direct efforts where they will be more effective, improving a program's "return," or success in reaching its goals. Treating all projects within a geography as equally important will result in a wide variability in performance and suboptimal outcomes.

New high-resolution geospatial datasets acquired through precision fieldwork and remote sensing provide a much more detailed understanding of the benefits that a competing proposed project might generate. These new data sources help reduce the risk that a project will deliver lower than expected results. Additionally, these analyses can be completed across entire landscapes, not only demonstrating an individual project's benefits but also providing context for how it compares with all other potential projects in the surrounding area.

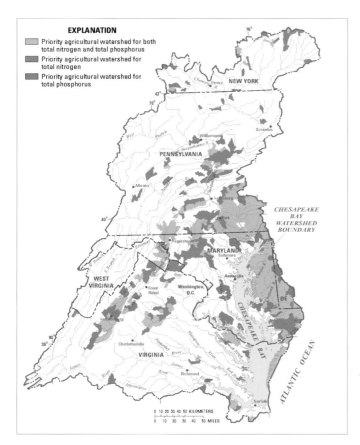

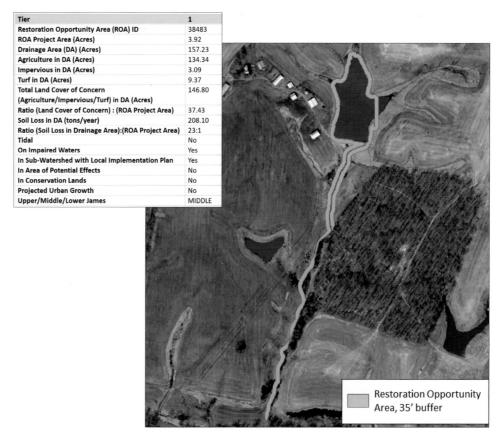

Tier	1
Restoration Opportunity Area (ROA) ID	38483
ROA Project Area (Acres)	3.92
Drainage Area (DA) (Acres)	157.23
Agriculture in DA (Acres)	134.34
Impervious in DA (Acres)	3.09
Turf in DA (Acres)	9.37
Total Land Cover of Concern (Agriculture/Impervious/Turf) in DA (Acres)	146.80
Ratio (Land Cover of Concern) : (ROA Project Area)	37.43
Soil Loss in DA (tons/year)	208.10
Ratio (Soil Loss in Drainage Area):(ROA Project Area)	23:1
Tidal	No
On Impaired Waters	Yes
In Sub-Watershed with Local Implementation Plan	Yes
In Area of Potential Effects	No
In Conservation Lands	No
Projected Urban Growth	No
Upper/Middle/Lower James	MIDDLE

High-resolution data and advances in modeling allow natural resource managers to move from evaluating restoration priorities at the watershed scale to identifying the potential benefits of individual projects within a larger landscape context. This level of detail reduces the variability of project performance and facilitates targeted investments to maximize programmatic outcomes.

Maintaining forward momentum through innovation

The CBP has effectively coordinated and managed Chesapeake Bay restoration for the last three decades based on strategies that use relatively coarse regional averages. Its ability to model the natural world at a finer scale was imperfect given previously available data and technology. Despite this limitation, the CBP made significant progress toward Watershed Agreement goals. Strategies for improving water quality must evolve going forward because much of the progress to date resulted from implementing the largest and most obvious ways to reduce pollution, such as upgrading wastewater treatment plants.

Future restoration efforts will increasingly focus on identifying and implementing high-performing projects across large landscapes. This transition has required the development of new models that use higher-resolution datasets that feature specific, localized priorities to deliver the greatest results.

Working in the places that matter most

The Susquehanna River provides about half the freshwater entering the Chesapeake Bay, including all of Pennsylvania's portion of the watershed. The Susquehanna is also responsible for about 40 percent of the total nitrogen load entering the bay. Much of the bad runoff comes from agricultural and stormwater sources. As such, implementing best management practices (BMPs) to reduce nutrient and sediment pollution from agriculture and storm runoff has become a top priority in the restoration of the bay.

Pennsylvania has identified several strategies to reduce the amount of nitrogen, phosphorus, and sediment entering the Susquehanna. Recently, the strategy of replanting forests along streams to create riparian buffers has gained favor. These trees act as a last line of defense against runoff and have been shown to be one of the most cost-effective ways to reduce pollution. Pennsylvania state agencies estimate that more than 95,000 acres of buffers must be planted to meet the Chesapeake Bay Agreement's goal of having 70 percent of riparian areas forested. The Chesapeake Conservancy is working with local partners to identify available planting areas that will treat as much of the 27,500 square miles of watershed as efficiently as possible.

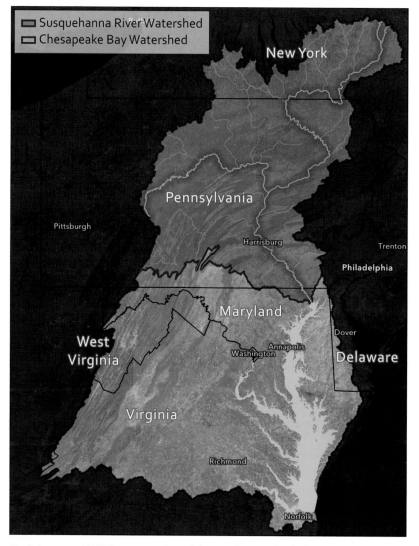

Draining over 27,500 square miles of the total watershed, the Susquehanna River spans parts of three states. Restoration work in the Susquehanna is critical to the overall health of the Chesapeake.

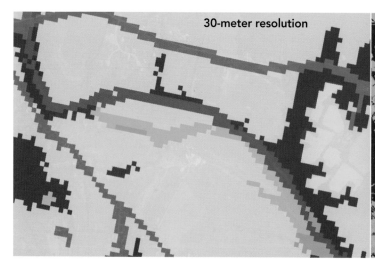

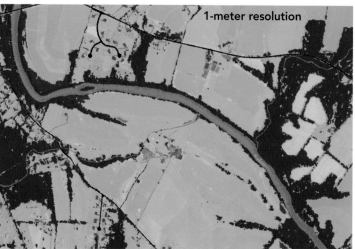

High-resolution data is transforming how managers model the landscape and identify priorities for restoration. New one-meter resolution land cover data shows a dramatically different picture of the landscape, and in particular, natural systems that have the potential to improve water quality, compared with the previously available 30-meter resolution data.

INNOVATING AT THE WATERSHED SCALE

Assessing individual opportunities for restoration across a watershed requires a different mind-set, and very different data than was used previously.

The Chesapeake Conservancy has worked with CBP and other partners since 2015 to create the geospatial data infrastructure for site-level planning. An endless variety of datasets can be incorporated into this type of prioritization, but the Conservancy found that two high-resolution datasets are critical for identifying potential opportunities: one-meter land cover data and lidar-based hydrology data.

Land cover data, created from one-meter resolution, four-band aerial imagery collected through the National Agricultural Imagery Program. This data is processed through an object-based image analysis workflow to segment and classify the imagery into discrete categories describing the landscape.

■ Land cover
■ Stream width
■ Flow path

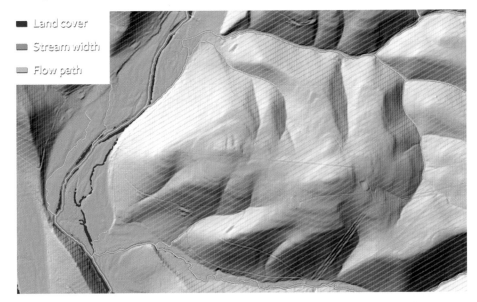

Hydrology and flow information, generated from lidar elevation data collected by states and the US Geological Survey (USGS). This data is processed through a series of steps to ensure hydrologic connectivity and to identify stream channels.

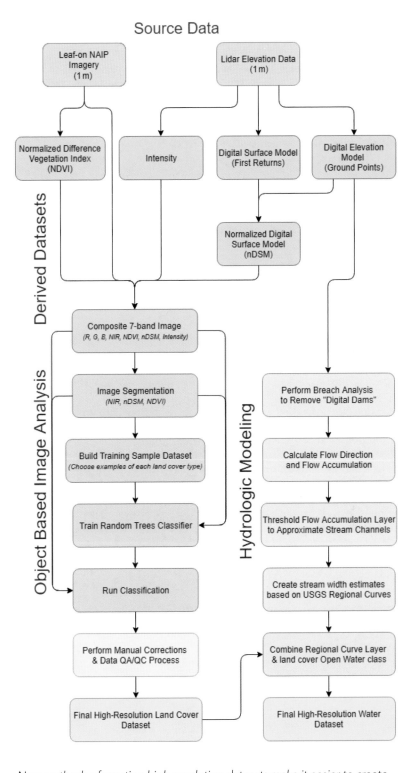

Source Data

Derived Datasets

Object Based Image Analysis

Hydrologic Modeling

New methods of creating high-resolution datasets make it easier to create comprehensive and highly accurate layers to model the natural world. This information opens new possibilites for landscape analysis and makes it possible to take into account the highly variable nature of the landscape. Conservation and restoration activities can have dramatically different results based on their locations, and these new datasets potentially ensure the implementation of the best projects.

A NEW FRONTIER FOR ENVIRONMENTAL MODELING

In 2016, CBP released the first Chesapeake Bay High-Resolution Land Cover Dataset, the result of an 18-month collaboration between the Chesapeake Conservancy, University of Vermont, and Worldview Solutions. Detailing more than 100,000 square miles, this 12-class, one-meter resolution dataset provides a consistent and accurate understanding of every county comprising the Chesapeake Bay watershed. With this new data, planners can create models that are accurate at high resolutions across large landscapes. This breakthrough is already providing new insights as to how small changes in project placement can dramatically influence performance. This previously unavailable level of detail has also enabled a variety of new analysis opportunities.

Managers can now approach environmental decisions in a far more informed way. Increasingly, decision makers can rely on data-driven processes to compare all options across the watershed and identify landscapes that offer the most potential "bang for the buck." Decision makers increasingly rely on data-driven processes

to compare options across a watershed and identify landscapes for conservation or restoration that offer the best cost-to-benefit ratio. The Chesapeake Bay High-Resolution Project showed that watershed-wide datasets can leverage economies of scale in data generation. And advances in distributive processing showed promise in speeding up complex workflows.

To support restoration efforts in the Susquehanna River and test the capabilities of a large-scale, distributed analysis, the Chesapeake Conservancy partnered with Microsoft Research and Esri to identify all the riparian restoration opportunities in the Susquehanna River watershed. This analysis leveraged the Chesapeake Bay High-Resolution Land Cover dataset and an enhanced water layer created from lidar flow paths that accurately represent the precise location of stream banks. The team used easily scalable, cloud-based servers to process almost 125 billion pixels in a matter of hours, which represents 5 percent of the time it would have taken to run the analysis on a local server. The analysis identified tens of thousands of acres of potentially restorable land in the Susquehanna River watershed.

Chesapeake Conservancy worked with Microsoft Research and Esri to adapt a traditional raster analysis workflow to operate on a cluster of virtual servers. This analysis identified restorable areas within 35 feet (11 meters) of streams and areas of concentrated flow in the Susquehanna River watershed. This distributed process ran approximately 20 times faster than if it was excuted on a set of local machines.

INFORMING LOCAL PRIORITIES

The Chesapeake Conservancy is working with local organizations across four counties in central Pennsylvania to further prioritize each of the 250,000 opportunities identified through the pilot study with Microsoft Research and Esri. Local partners are using this information to inform local restoration initiatives for a portion of the Susquehanna River.

This process assesses the additional benefits a buffer restoration project could provide, such as increasing habitat for threatened species, improving climate resilience, and expanding regional ecosystem connectivity. An increased understanding of the benefits of an expanded riparian buffer can help managers determine the relative importance of a potential project beyond water-quality improvements. Each project receives a "site score" detailing its benefits and a "designation score" based on whether it is located in a priority watershed. Partners can use the composite of these two values to evaluate projects and better understand how well the project meets organizational goals.

In its first year of use, ten projects were evaluated using the conservancy's analysis prior to being implemented, including a project that was ranked highest of any potential project in all four counties.

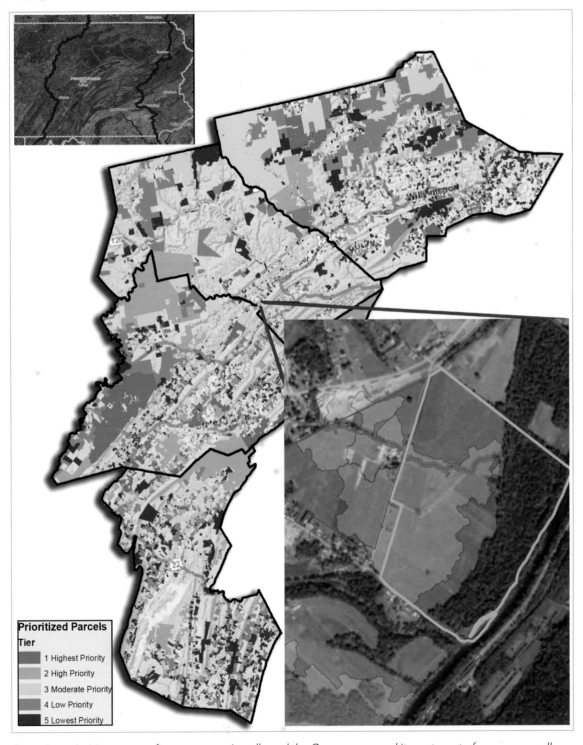

Prioritized Parcels

Tier

- 1 Highest Priority
- 2 High Priority
- 3 Moderate Priority
- 4 Low Priority
- 5 Lowest Priority

Riparian buffer restoration projects, such as this site in Centre County, Pennsylvania, can significantly reduce the amount of nutrient and sediment pollution entering local streams and, eventually, the Chesapeake Bay.

Assessing priorities across a four-county region allowed the Conservancy and its partners to focus on a small subset of high-performing projects. Using this approach, the Conservancy was able to narrow the number of opportunities from over 200,000 to the top 278 projects, or 0.1 percent. These projects, on average, treat the runoff of 40 acres of upland for every acre of restoration, ten times what is used in CBP's modeling.

MOVING THE DIAL

Prior to having this prioritization data, restoration projects were often selected based solely on the size of the project or, in some cases, simply because there was a willing landowner. This approach would allow partners to meet effort-based goals but has a high potential for low-performing projects to be implemented.

When the Conservancy analyzed the conservation scores from the first year's projects, generated by evaluating each project's anticipated performance and location in impaired watersheds, there was a wide variety of scores, including some extremely low-value projects. Partners reflected on how access could benefit their programs and made a commitment to use the planning data in their community outreach and recommend only higher-ranking projects for funding.

Riparian buffers act as a last line of defense for stopping pollution before it enters streams, and eventually the Chesapeake Bay. Ideally, these trees will be planted where the landscape converges so they intercept as much runoff from upslope as possible. Traditionally, models have not had the detail to measure the actual amount of area being filtered by a restoration project, so an average value was used. With the detail in our elevation data, however, we can specify the exact landscape being treated by each project and incentivize higher-performing projects.

The projects selected in our second year of funding should filter pollution from 11 times the amount of upland area that the previous models—based on a fixed upland value—would have treated. The treated area represented an 18 percent increase from the previous year's average. This four-county pilot region demonstrates the potential for new models to have an outsized impact on the performance of restoration work throughout the Chesapeake Bay and around the world.

The Chesapeake Conservancy is now working with CBP and other nonprofit and academic partners to build the datasets needed to apply models across the watershed similar to the ones developed through this project.

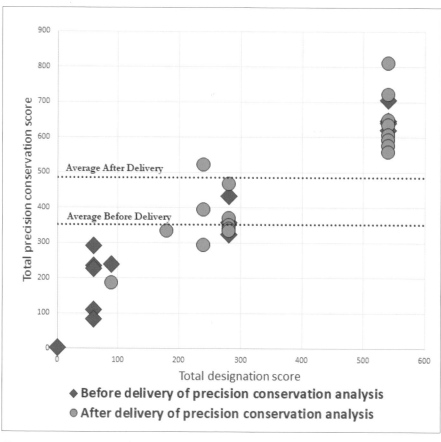

Conservationists now evaluate projects within a system that compares each proposal not only within the subset of projects that were submitted, but also within the larger context of the landscape. Projects implemented prior to using the prioritization system (orange diamonds) had a wide variety of performance, including one project that provided little environmental value. In the second year, projects that had previously been identified and ranked high were advanced due to their performance (blue circles), showing how a small change in management can lead to a dramatic improvement in performance.

Volunteers plant native tree species along Elk Creek in Centre County, Pennsylvania. These trees will help reduce pollution by stabilizing the stream's banks and filtering out nutrients and sediment that flow downhill from the surrounding landscape.

LEVERAGING THE POWER OF AI

In 2017, Microsoft Research launched its AI for Earth initiative to focus the capacity of artificial intelligence on improving environmental decision making around the world. The conservancy joined the AI for Earth initiative to focus on the use of artificial intelligence in the bay's restoration. Specifically, the conservancy initiated a project to rapidly update its high-resolution land cover data as new imagery becomes available.

One of the challenges facing any approach that uses AI is the need for highly accurate training data that represents a variety of conditions. The Chesapeake Bay High-Resolution Land Cover Dataset represented an ideal training set because of its geographic scope, the variety of landscapes it includes, and its high degree of accuracy. This system uses Microsoft's Cognitive Toolkit, an architecture designed to easily train deep learning algorithms. The system also uses imagery from the National Agriculture Imagery Program (NAIP) stored on Microsoft's Azure Cloud and a new platform called Project Brainwave that integrates Intel's FPGAs—field programmable gate arrays—to increase the throughput capacity of Azure while providing increased flexibility as the system scales.

A deep learning algorithm analyzed the existing land cover data in combination with the four-band NAIP imagery to create a simplified four-class land cover dataset in the Chesapeake. With the algorithm trained, two datasets were created, one based on 2013/14 NAIP imagery (the source data for the Chesapeake Bay High-Resolution Land Cover) and one based on 2016/17 NAIP imagery. This tested the platform's ability to create a classification based on imagery that was different from the data on which the system was trained. Both datasets ran successfully, and the conservancy is comparing the two time periods to identify areas that have experienced significant levels of change over the three-year period. This analysis will inform future updates to the Chesapeake Bay High-Resolution Land Cover dataset.

The conservancy's work had moved into geographies that lacked high-resolution land cover data, including landscapes where lidar data was not available. As a result, the partners expanded the system so that it could create high-resolution land cover data anywhere in the United States.

Project Brainwave significantly accelerated the processing capabilities of this large-scale analysis and completed the classification in a little over 10 minutes at a cost of $42. The classification does a good job of detailing landscapes that look visually similar to the Chesapeake, such as the Great Lakes or Pacific Northwest, but struggled in arid landscapes. Although it doesn't completely replace our previous workflow yet, this system illustrates how AI has the potential to quickly create high-quality data.

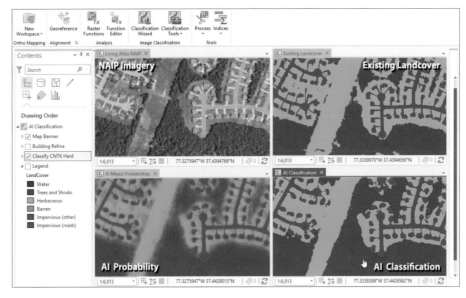

The land cover algorithm the Conservancy developed with Microsoft and Esri leverages NAIP imagery (top left) and the Conservancy's Chesapeake High Resolution Land Cover (top right) to predict the likelihood of each pixel matching one of four classes (bottom left). Based on this probability layer, the algorithm chooses the most likely class for the final classification (bottom right).

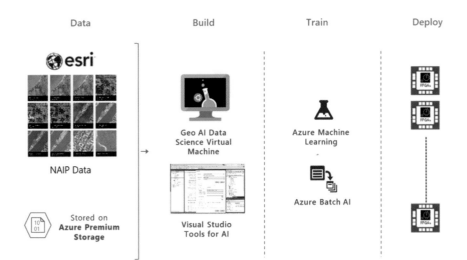

200M Images, 20TB
Land cover mapping for the whole of US in
10+ minutes

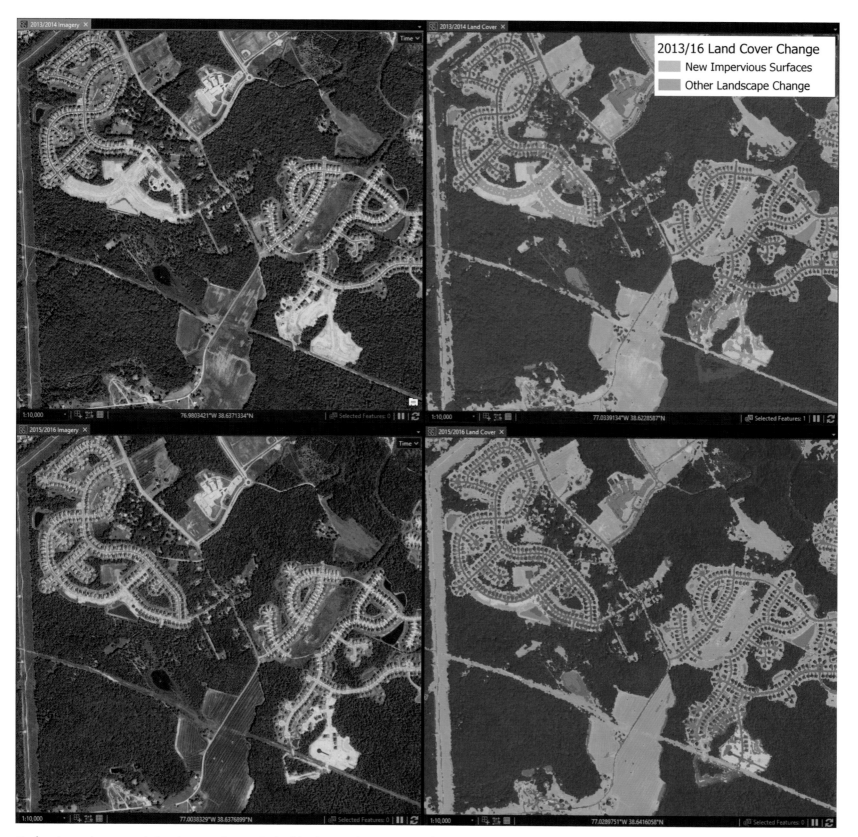

Performing a change analysis using two time periods of land cover data in Microsoft's Azure Cloud allowed the Chesapeake Conservancy to quickly identify priority landscapes in updating the Chesapeake Bay High-Resolution Land Cover dataset.

AI IS CHANGING THE WAY WE THINK

Access to current information is critical for environmental management so that conservation and restoration partners can accurately understand the world around them. Ensuring data accuracy can be costly and can leave less funding for program implementation.

AI-generated datasets can save money and time as the conservancy and its partners update critical landscape data. As these systems improve, updated landscape data could be automatically processed each time a satellite passes over Earth.

This technology will help partners increasingly focus their efforts on the areas that are experiencing the most change. Instead of noticing the impacts to an ecosystem by observing a decline in its health years after the fact, AI-based systems will provide an early-warning system as critical landscape thresholds, such as the percentage of impervious surfaces in a watershed, are exceeded.

Updates to the Chesapeake Bay land cover are already being informed by a rapid-change analysis leveraging the two time periods of data created in Azure. This analysis is helping the conservancy identify where hotspots of change have occurred during the last three years and an updated land cover dataset is needed to better reflect the landscape.

Moving to other landscapes

Managing water quality is a priority for all landscapes. Chesapeake Conservancy's approach to improving conservation and restoration design has the potential to improve decision making and ensure that projects are as efficient as possible.

One of the greatest limiting factors to expanding these analyses, however, is the time and cost of creating the base datasets needed to run more advanced modeling. Integrating AI-based workflows can remove these barriers and expand precision planning capabilities to new landscapes.

In 2017, the Iowa Agricultural Water Alliance asked the conservancy about applying the Susquehanna pilot program to the Black Hawk Creek headwaters, a 160-square-mile subwatershed in central Iowa. This project was part of a larger initiative working to identify ways to accelerate the process of developing more than 1,600 watershed management plans spanning the state.

This project demonstrated the ability of the Azure-based land cover workflow to accurately classify a landscape outside the Chesapeake. More importantly, the workflow allows projects to be completed within a tight budget and extends planning capabilities to landscapes where generating new data would be impossible otherwise.

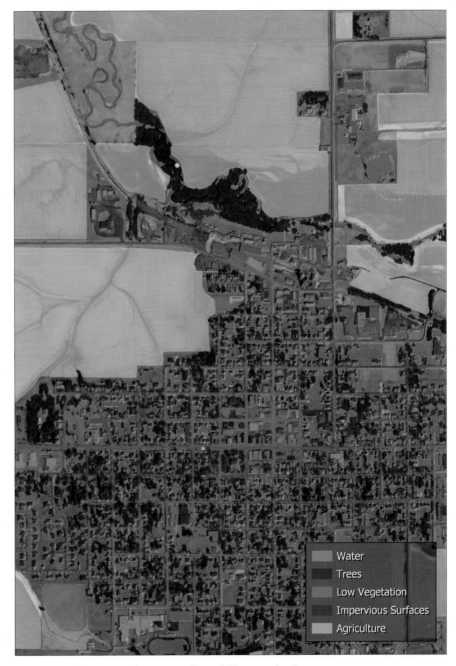

Land cover data created in Azure allowed Chesapeake Conservancy to assess restoration opportunites on agricultural lands along Black Hawk Creek in Iowa. This project would not have been feasible given the available budget if the data had to be generated manually.

WHERE WE GO FROM HERE

Chesapeake Conservancy is expanding its portfolio of work and applying its expertise outside the Chesapeake Bay watershed to work in landscapes across the United States and, increasingly, around the world. The Conservancy's approach to precision conservation has the potential to improve environmental outcomes for both water quality and water quantity management as well as a host of other environmental issues such as habitat conservation and climate adaptation. However, moving toward a performance-based system of management requires access to high-quality, and often high-resolution, datasets to be as effective as it has been in the Chesapeake.

To help address these management challenges in varied landscapes, the Conservancy is continuing to work with Microsoft to refine the land cover classification algorithms to improve the accuracy of the output, make it easier to retrain the algorithms in new landscapes, and explore the potential to add additional classes. This work holds a tremendous amount of promise to rapidly produce usable datasets that can easily be shared with partners.

Most importantly, this system is enhancing the Conservancy's capacity to generate repeat datasets whenever new imagery becomes available. With an influx of new high-resolution satellite datasets, access to this platform will allow us to create a "living" dashboard of the watershed that reflects conditions as soon as they change instead of only incorporating changes years after it has occurred. Having this dynamic system will provide managers with the information they need to customize and tailor solutions where they will be most effective.

Clear waters in the lower Chesapeake Bay deep in Virginia.

PART 4
TRAINING FUTURE GENERATIONS OF SCIENTISTS

Read the inspiring stories of students doing science in the greatest classroom of them all—outdoors in the field. From drone and satellite imagery to GPS and field observation, students of anthropology, conservation biology, forestry, glaciology, and many other disciplines are learning to use traditional and cutting-edge tools, technologies, and methods in their research. These students and their instructors and mentors show us "what's next" in training this next generation of scientists.

A GLACIER IN RETREAT

Wyoming is the third-most glaciated state in the United States after Alaska and Washington. The quest to measure the extent of ice retreat and predict the implications of losing the state's 80-plus glaciers has led a multidisciplinary research team to the Dinwoody Glacier at the base of Gannett Peak—Wyoming's tallest mountain.

By Jacki Klancher, Todd Guenther, and Darran Wells,
Central Wyoming College

Current terminus
or toe of glacier

Terminal moraine from
Little Ice Age glaciation

Once covered with ice down to the edge of the gray, glacial moraine, the Dinwoody Glacier
is in rapid retreat, having lost half its area and even more of its mass in the past 50 years.

—3D view generated from ArcGIS Pro with Imagery basemap

CHASING THE ICE

On a clear August day, student researchers pull a yellow sled across a snow-covered glacier in the Wind River Range of Wyoming. The group follows the footsteps of a leader who trudges 100 feet ahead with a handheld GPS unit. A half mile away, another crew flies a large kite with a GoPro attached. Farther down the snow and ice, another parka-clad contingent gathers clean snow in clear plastic bags that they'll melt, filter, and catalog later at basecamp. Far below, past the end of the terminal moraine, yet another group photographs a gridded archaeological site.

Stumbling upon this close-knit team from Central Wyoming College (CWC), one might conclude they're going about a collection of random activities. In fact, these 18- to 21-year-old students are part of the highly organized Interdisciplinary Climate Change Expedition (ICCE). They have a focused mission to determine changes in the ice depth and extent of the Dinwoody Glacier at the base of Gannett Peak (Wyoming's highest point) and study the implications of further ice retreat on the water resources and the immediate and front country environments. Using a variety of geospatially enabled technologies, such as GPR (ground-penetrating radar), KAP (kite aerial photography), handheld GPS (global positioning systems), and other instruments, the expedition has returned to the glacier for six consecutive years.

Dinwoody Glacier is located in the state of Wyoming on the east side of the Continental Divide in the Wind River Range.

"This is not an easy place to do research," says ICCE leader Jacki Klancher, an environmental science professor. "We're at 11,000 feet. The air is thin. We're 24 miles from the nearest road. It took three days to hike in. We've had to pack in and use mountaineering gear to stay safe on the ice. But all this is worth the effort because we're collecting reliable data about the critical issue of glacial ice loss that affects everyone in the state—whether they know it or not. If most people are not even aware of the presence of these glaciers, they are most likely also unaware of the implications of losing them. We can now say with high certainty this glacier lost 13 meters of ice thickness between 2006 and 2016. It's only 50–100 meters thick right now in the deepest areas, so that's a significant drop."

In addition to the 50 pounds or so that students carried on their backs, the team also hired a packhorse team to bring in another 900 pounds of science gear and additional food rations for this year's 16-day expedition. This expedition is facilitated by its proximity to the Wind River Range and close ties with CWC's Outdoor Education degree program and NOLS, the respected National Outdoor Leadership School headquartered just blocks from CWC's Lander campus. One unique aspect of this program is that the students have the opportunity to join a research expedition as freshmen and sophomores at a community college. At many schools, only upper-division and graduate students get to participate in rigorous field research. Once they're down from the mountain, many of these young adventurers will embark on their first GIS courses, some eventually choosing scientific careers leading to the University of Wyoming and beyond.

Scope of research

The goal of the ICCE program research is to collect quantifiable data about glacial ice in Wyoming that can be analyzed to assess the impact of a warmer, drier climate on the state's ecology and hydrology. This research includes measuring ice thickness and its extent over time and the presence of micro-particulates in the snow. Researchers also study the dynamics of hydrologic flows, hunt for biological organisms, and search for prehistoric evidence of human settlement in the high mountains.

In its six-year history, the expedition has focused on specific areas of GIS-underpinned research:

- Measuring ice thickness using GPR
- Analyzing comparative image analysis
- Measuring black carbon particulates
- Measuring water flow from glacial melt
- Detailing site orthophotography with KAP
- Documenting human settlement (archaeology)

"GIS and geospatial thinking are woven into everything we do," Klancher says. "Back in the lab, we use ArcMap™ and now ArcGIS Pro for analysis. We share data on ArcGIS Online; we rely on the high-resolution Sentinel imagery and many other layers like weather and precipitation feeds found in the ArcGIS® Living Atlas of the World, and perhaps most importantly, we publish the results of our experiences online via story maps. Sadly, the majority of Wyoming residents are not aware that all of our state's glaciers are melting away. We keep looking for compelling ways to engage with them and anyone else who will listen."

A REMOTE RESEARCH STATION

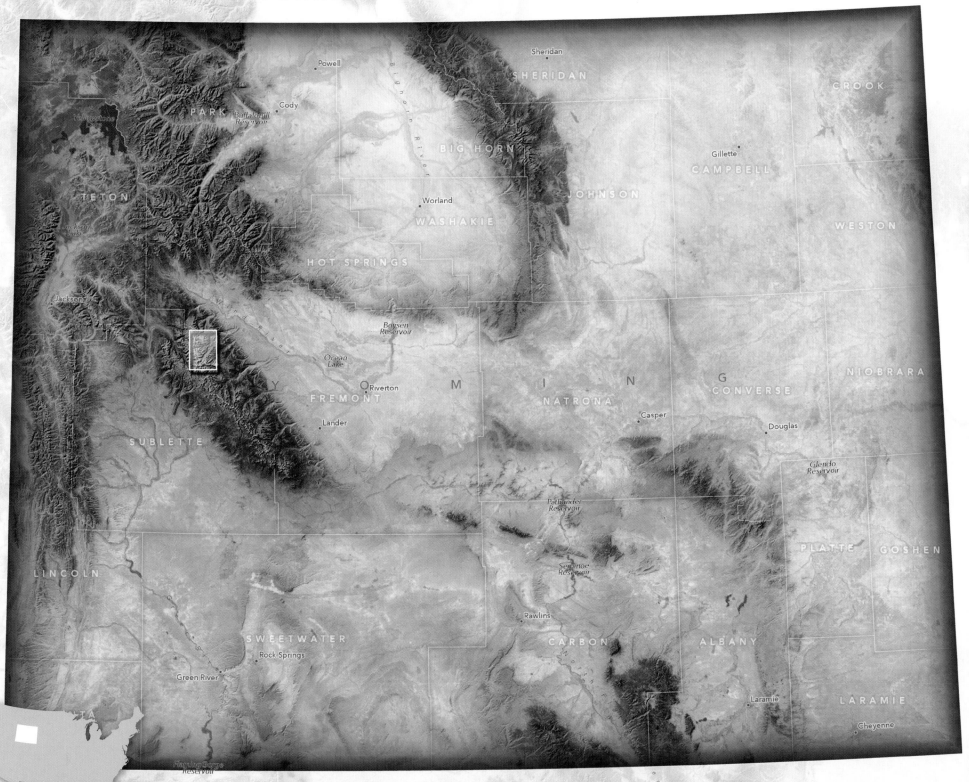

More than 80 glaciers exist in the mountains of Wyoming. The Wind River Range, part of the southern extension of the Rocky Mountains, is located in the west-central region of the state. The rugged wilderness is part of the Greater Yellowstone Ecosystem. This remote area is home to the majority of the state's glaciers, what hydrologists think of as frozen reservoirs.

ON THIN ICE

The original goal in the first year of the expedition was to physically measure the thickness of ice along a specific transect (a sample line across the glacier). A pioneering 1970s expedition mapped the transect originally, so by duplicating the same path, the team collected valuable comparative data.

GPR is a geophysical method that uses radar pulses to image the subsurface beneath a layer of ice and snow. Considered environmentally safe, the method sends out pulses in the radio band of the electromagnetic spectrum and detects the reflected signals from bedrock beneath the ice.

The 2017 team followed the same transect that the previous years' expeditions used (which was also the original one). To ensure they followed the same transect, team members downloaded the waypoints onto a GPS unit that the leader used to trace the path for the GPR sled.

GPR field team lead Darran Wells points out that this is a long-term effort. "It will take several more years of accumulated data before we can definitively release conclusive measurements of ice loss over time." GPS measurements of the Dinwoody Glacier terminus from 2017 provided misleading estimates of the glacier's position, as it was difficult to separate glacial ice from the snowpack. The Forest Service has recognized the value of all the data and granted researchers a five-year permit extension to better determine if and when the glaciers will disappear.

Wearing crampons and helmets, the team pulls a Noggin GPR unit with a 100-MHz antenna to collect subsurface data from the glacier.

The red line is the path followed by the 2016 and 2017 transects. The green line marks the end of the ice extent as measured by a team walking and collecting waypoints with a GPS unit. The scale is deceiving; the glacier measures more than a mile from side to side.

Leader follows previous
year's transect with a GPS

Transect line

GPR sled team

A high buttress dwarfs a small team of scientists collecting geospatially
referenced depth data along a traverse of the Dinwoody Glacier.

GPR METHODS AND RESULTS

The GPR team had two main objectives: first, determine the practicality of using this technology for analyzing ice depth of a backcountry high-alpine glacier, and second (if researchers could collect good data showing depth to bedrock), create a baseline measurement of ice depth for future comparison.

Few researchers have used GPR in such remote and largely inaccessible alpine glacier studies. Using the technology in deep wilderness is often impractical because of the unwieldy shape and weight of the GPR antennae. Despite these limitations, GPR may prove to be a useful tool for in situ analysis of ice depth for mountain glaciers.

A Noggin 100-MHz antenna was used to collect subsurface data from the Dinwoody Glacier. The team used Universal Transverse Mercator (UTM) coordinates from a 2006 study to map and retrace a previously measured transect. Every tenth coordinate listed in the 2006 study was entered into a Juno SB GPS unit, which was used to direct the GPR team along the transect. Beginning at the southeast corner of the Dinwoody Glacier and contouring a distance of 1,563 meters around to the northwest edge, the GPR detected bedrock for a majority of the transect.

Data collection involved one team member being tethered to the front of the unit while two other members kept the GPR unit from sliding out of control down the icy slope. Each team member wore a helmet, harness, and crampons. Care had to be taken not to allow metal crampons and ice axes to interfere with GPR data collection. Data collection was slow and tedious. Frequent exposed rock made movement especially difficult, and much of the transect forced the GPR team to navigate steeply angled terrain. The team collected GPS points using the Noggin unit while also recording the transect. Before starting the research transect, a 535-meter test was run on the glacier to ensure the GPR unit was accurately collecting data.

The ICCE team demonstrated that GPR provides highly usable and effective field data and can more accurately detect the thickness of ice than the analysis of remotely sensed images alone. Combined with remote sensing, GPR can provide a strong field tool to provide a more comprehensive dataset to describe glacial conditions.

With months of preparation invested and much at stake, the GPR team packs in a ruggedized PC laptop to review and validate their hard-earned data in the field.

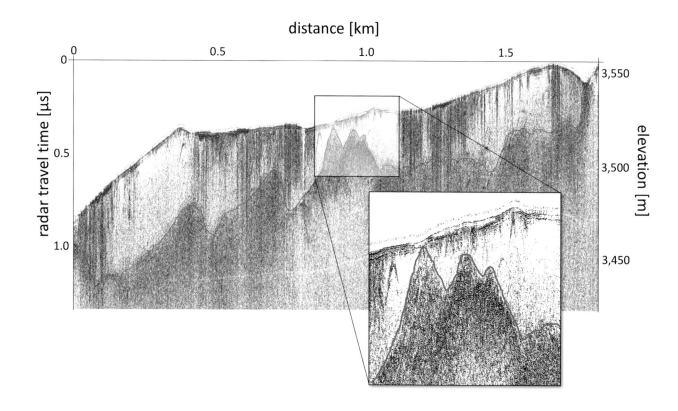

distance [km]

radar travel time [µs]

elevation [m]

The research results indicate that the Dinwoody Glacier continues to recede—both the terminus itself and the overall volume. GPR data obtained in 2016 and 2017 shows the thickness of the ice throughout the length of the transect. This data reveals a decrease in ice depth of approximately 13 meters since 2006, slightly more than 40 feet—the height of a four-story building. GPR analysis is performed under the guidance of Dr. Andrew Parsekian at the University of Wyoming.

The 2017 expedition found depth to bedrock (ice thickness) ranging from under 10 meters up to about 60 meters of depth in the thickest sections.

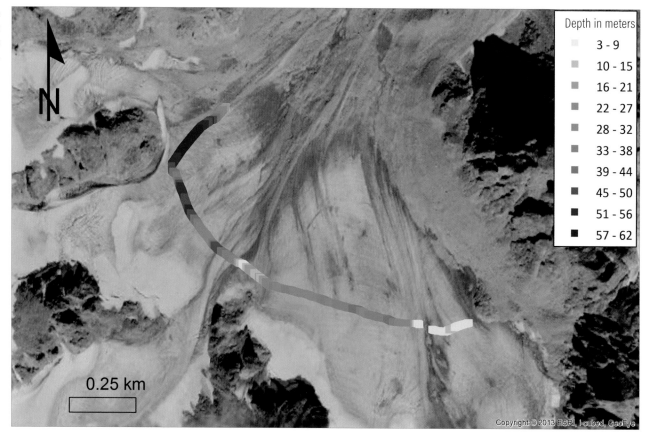

Depth in meters

	3 - 9
	10 - 15
	16 - 21
	22 - 27
	28 - 32
	33 - 38
	39 - 44
	45 - 50
	51 - 56
	57 - 62

0.25 km

Copyright © 2013 ESRI, I-cubed, GeoEye

KITE AERIAL PHOTOGRAPHY

While the GPR team measured the thickness of the ice, a new endeavor in 2017 attempted to capture detailed imagery and 3D terrain data using a kite-mounted camera. Whereas a drone would have been the ideal solution, federally designated wilderness rules strictly prohibit powered vehicles in the area.

Kite aerial photography (KAP) is a type of imaging in which a camera is lifted using a kite and is triggered either remotely or automatically to take aerial photographs. The camera rigs can range from the extremely simple, consisting of a trigger mechanism with a disposable camera, to complex apparatus using radio control and digital cameras. On some occasions it can be a good alternative to other forms of aerial photography.

Part of the difficulty in bringing new sensor technology such as the kite rig into the project involved some initial growing pains and a crash course in the power of Mother Nature to upset the best-laid plans. During the 2017 expedition, the team initially faced doldrums with hardly enough wind to fly the kite, and then, with little warning, intense 60-mph gusts that ultimately destroyed the kite after collecting less than the expected amount of data. Upcoming expeditions will take these lessons to heart and try again.

A Picavet suspension, named after its French inventor Pierre Picavet, consists of a rigid cross suspended below the kite line from two points. A single line is threaded several times between the points of the cross and the points of attachment to the kite line, and the rig is attached to the cross. The Picavet line runs through eye hooks or small pulleys so that the weight of the rig causes it to settle naturally into a level position. The dimensions and shape of the Picavet have been adapted many times in attempts to increase stability and improve portability.

GoPro camera hanging from
a Picavet suspension

The KAP team walks the ice, traversing as much ground as possible. Drones could have captured the scene in much less time, but the Dinwoody Glacier is located in the federally designated Fitzpatrick Wilderness Area with strict prohibitions on both motorized and mechanized vehicles—including drones. That makes kites an attractive alternative.

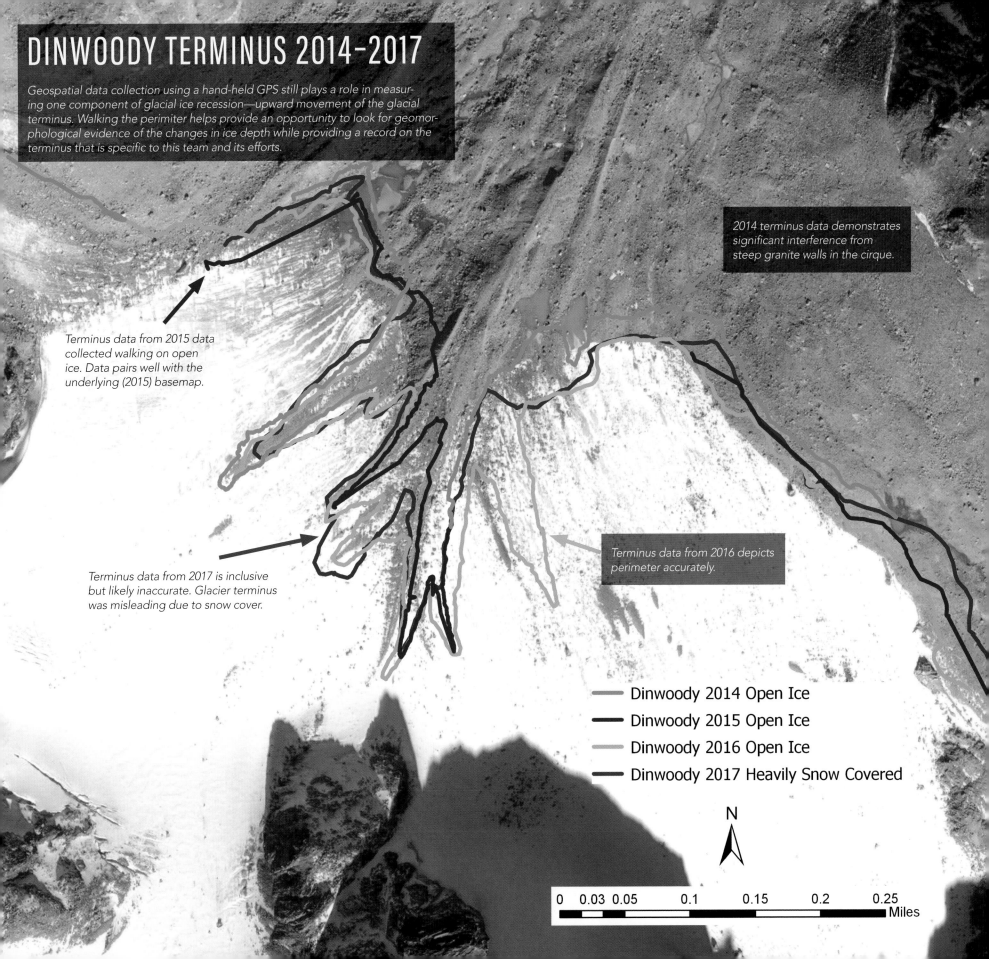

DINWOODY TERMINUS 2014-2017

Geospatial data collection using a hand-held GPS still plays a role in measuring one component of glacial ice recession—upward movement of the glacial terminus. Walking the perimeter helps provide an opportunity to look for geomorphological evidence of the changes in ice depth while providing a record on the terminus that is specific to this team and its efforts.

2014 terminus data demonstrates significant interference from steep granite walls in the cirque.

Terminus data from 2015 data collected walking on open ice. Data pairs well with the underlying (2015) basemap.

Terminus data from 2016 depicts perimeter accurately.

Terminus data from 2017 is inclusive but likely inaccurate. Glacier terminus was misleading due to snow cover.

Dinwoody 2014 Open Ice

Dinwoody 2015 Open Ice

Dinwoody 2016 Open Ice

Dinwoody 2017 Heavily Snow Covered

N

0	0.03	0.05	0.1	0.15	0.2	0.25

Miles

BLACK CARBON AND HYDRO

In addition to the primary mission of recording ice-depth measurements, ICCE has also extended its research scope into the nature of the snow itself and the flow of meltwater from the glacier.

Black carbon

Researchers sought to understand whether the presence of light-absorbing black carbon particulate matter contributed to the loss of glacial ice and increased snow melt. Any dark particulate matter in the ice decreases snow albedo (degree of reflectivity off the snow or ice) and leads to increased absorption of infrared energy (heat). The naked eye cannot see these small particles, but researchers can filter the particles from melted snow for lab analysis. Black carbon particulate results from the incomplete combustion of organic matter. The carbon is emitted into the air from various sources, but most notably forest fires, agricultural burning, and the combustion of fossil fuels. When deposited in the snowpack, this particulate can considerably accelerate the rate of snow and ice melting. At values of 30 nanograms of black carbon per gram of water (ng/g)—typical of North American snow samples—the albedo is decreased between 2.5 percent and 6 percent.

The average values for black carbon particulate matter in the Dinwoody Cirque in 2018 range from 3 to 121 ng/g of effective black carbon, with an average from all 22 collection sites of 57 ng/g. (A measurement of 191 was recorded in 2016, attributable to the massive Lava Mountain forest fire that burned out of control nearby that summer.)

Dr. Carl Schmitt, who designed a cost-effective way to analyze samples, oversees the methods of black carbon data collection and analysis. Schmitt is credited with providing ICCE students with detailed instruction on prescribed field methods and completing the analysis of the black carbon field samples in 2018.

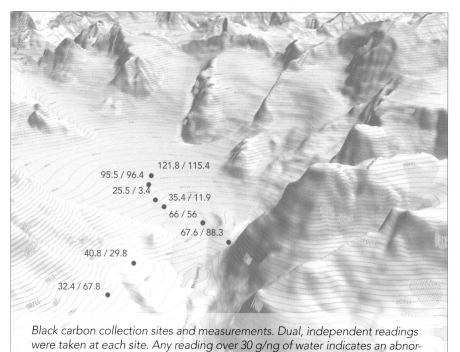

95.5 / 96.4
121.8 / 115.4
25.5 / 3.4
35.4 / 11.9
66 / 56
67.6 / 88.3
40.8 / 29.8
32.4 / 67.8

Black carbon collection sites and measurements. Dual, independent readings were taken at each site. Any reading over 30 g/ng of water indicates an abnormal level of black carbon in the snow pack.

Hydrology

Stream volume measurements were recorded at five sites below the terminus of the Dinwoody Glacier. The goal was to determine what percentage of overall downstream volume in Dinwoody Creek (as measured by a downstream US Geological Society [USGS] gauging station) is comprised of water emerging from snow and ice melt from the hydrologically linked Dinwoody and Gooseneck Glaciers. When the volume of glacial meltwater was measured against the total downstream volume of Dinwoody Creek, the contributing glacial meltwater was estimated at 28 percent for 2016 and 7 percent for 2017. Given that the snowpack in 2016 was very low and August glacial ice highly exposed to sunlight, this value seems plausible. In contrast, 2017's modest contribution of 7 percent is believed linked to greater protection from the higher winter snowpack and lower ambient temperatures. Earlier assessments made by Pochop et al. in 1990 suggested the overall contribution of the Dinwoody, plus the streamflow from the Gannett Glacier (the largest glacier in the Wind River Range), comprised 13 percent of total stream flow in Dinwoody Creek. Collectively, these studies both target glacial meltwater as a significant contributor to the surface water volume of Dinwoody Creek.

In the future, annual meltwater discharge values at the Dinwoody terminus will be compared across years to establish a quantitative estimate of the percentage of downstream surface water being fed by the Dinwoody and Gooseneck Glaciers during the month of August.

Measuring flow of Dinwoody Creek

Student hydrologists collect clean snow samples from beneath the surface. The material is bagged and carefully labeled and then filtered in camp.

HISTORICAL MAPS AND IMAGERY TELL A TALE

In 1951, a graduate student from the University of Iowa, Mark F. Meier, chose the region as the subject of his master of science degree. His detailed maps and photos serve as a benchmark for modern scientists. Additionally, a number of historically important and detailed photographs of the Dinwoody Glacier exist, dating to the 1930s. The images do not provide quantifiable measurements of ice depth, but taken together, the historic images create an informative visual record. Meier's hand-drawn 1950 map was imported into ArcGIS Pro and georeferenced to recent-year satellite imagery. Tracing the extent of the 1950 glacier and placing it on the most recent satellite image is another effective method of visualizing the significant retreat of this glacier in less than 70 years.

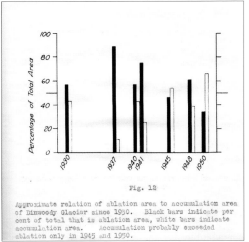

Fig. 12

Approximate relation of ablation area to accumulation area of Dinwoody Glacier since 1950. Black bars indicate per cent of total that is ablation area, white bars indicate accumulation area. Accumulation probably exceeded ablation only in 1945 and 1950.

The problem of decreasing glacial thickness is not new. Measurements made between 1930 and 1950 already showed significant ablation of the glacial ice. The rate of change has only accelerated since that time.

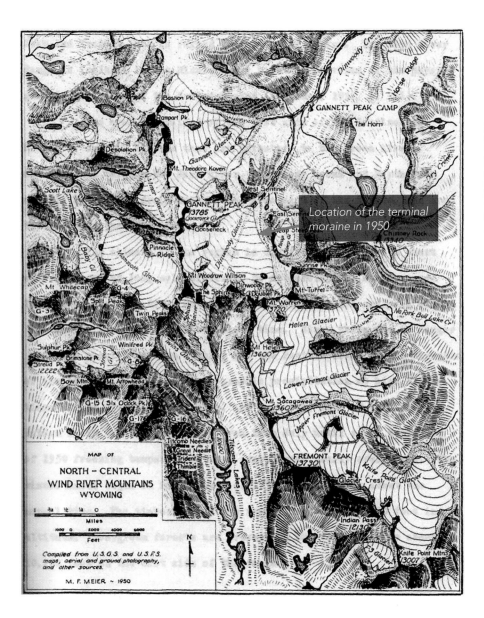

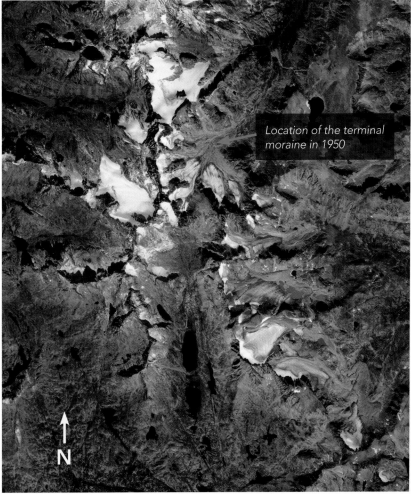

THE FIRST MOUNTAINEERS

The most recent prehistoric inhabitants of the high country in Wyoming, Montana, and Idaho called themselves *Tukudeka*—the Sheep Eaters. These people created much of the archaeological record studied by CWC students. One of many branches of the Shoshone Tribe, they lived in montane regions, and mountain sheep were a prominent part of their diet. Depending on where they lived, other Shoshone groups were called Buffalo Eaters, Root Eaters, Fish Eaters, and so on. The Tukudeka were also called Mountain Shoshone, or *Toyahini*, "the mountaineers." Their encyclopedic geographical knowledge was displayed to the earliest white trappers who asked them to draw charcoal maps on "white elk skin." We don't have names for the people who lived in the alpine reaches of the Wind River Range before the Sheep Eaters, but archaeological evidence proves they were there since the dawn of American humanity. Nevertheless, many people, public and professionals, question prehistoric human use of the high country.

"I don't know why people can't believe we were up here," mused one Shoshone archaeology student at CWC while working on an 11,000-year-old campsite at the top end of the Dinwoody Canyon. "Grandma was a Sheep Eater, and she told all us kids all about it by the time we were eight years old."

Every prehistoric period is represented in the villages and campsites that cover the ridges and valleys of the Wind River Mountains. People have ventured into the high country for generations and for thousands of years, attracted by food, beauty, and spiritual places where water from the ice brings life to the valleys below the glaciers. The Eastern Shoshone—descendants of those prehistoric people—still live in the foothills on the Wind River Indian Reservation, along with members of the Northern Arapaho Tribe. Many of these descendants continue to venture into their mountainous ancestral homeland to hunt, gather firewood, and restore their spirits in the sanctity and beauty of the high country.

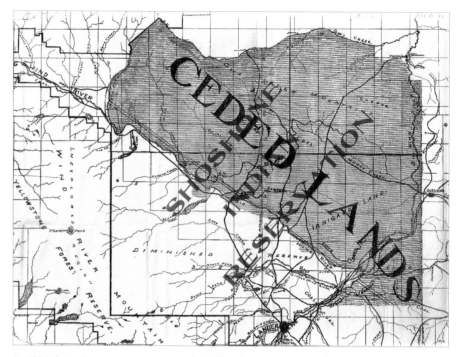

In 1906 the US government coerced the Shoshones and Arapahos into giving up more than half of the Wind River Reservation for white settlement. Vast irrigation projects were planned and built to support white-owned farms on former reservation land north of the Wind River. This map also shows the major rivers and streams that flow out of the mountains and across the southern portion of the reservation.

This photo shows the famous Shoshone Chief Washakie's camp in the south end of the Wind River Mountain chain during the late 1870s. Many Sheep Eater families joined Washakie's camp after they were forced out of the mountains and onto the reservation.

Northern Arapaho leader Goes-In-Lodge overlooks a mixed Shoshone and Arapaho camp on a bank of the South Fork of the Little Wind River where it flows out of the mountains. Archaeological evidence documents that people attracted by beauty, shelter, and water have camped in this spot for 10,000 years—since the end of the Pleistocene Ice Age. This photo shows a 1920s Hollywood film set used to tell stories of interracial friendship, warfare, and the establishment of the Wind River Reservation in 1868. Chief Washakie originally selected the location partly because of the reliable water that flows down from the glaciers and sustains life in the valley below.

ANTHROPOLOGY

By CWC Archaeology Professor Todd Guenther and CWC student Mara Gans

Approximately 12,000 years ago, humans from the north first entered Wyoming in the middle of the North American continent. These cold-adapted hunters possessed a thorough understanding of their local geography. We tend to think their geographic information was simple: maps scratched in the sand or on a cliff face, charcoal lines on a hide, or old stories passed down from elders. Modern neuroscience suggests that their innate wayfinding skills and locational knowledge is hardwired into the human brain. Their survival depended on a deep understanding, both learned and intuitive, of the relationships between elevation, latitude and longitude, geologic and vegetation zones, animal migration routes, and myriad other variables of the natural world.

Their campsites and kill sites are scattered across the North American landscape, but those earliest people did not stay in the warm valleys. They hiked to the highest reaches of the most glaciated mountains in the central Rockies. Clovis and Goshen spear points found just below the toe of the Dinwoody Glacier demonstrate that the earliest well-documented humans in North America journeyed up to the terminus of glacial ice. Like the Incas in Peru, these people followed the ice, as did their descendants. Their campsites and religious sites trace the retreat of the glaciers up the canyons at the end of the Pleistocene Epoch and up to the present. Prehistoric peoples lived in a dynamic world. Climate changed, ecozones changed, flora and fauna changed, and people capitalized on those changes to feed themselves.

Modern archaeology and environmental science students at CWC use advanced geospatial technologies to locate, record, describe, and interpret past human activities and impacts. Conventional wisdom long held that the thin air, cold temperatures, and short summer seasons in the high alpine rendered the region uninhabitable by humans. Recent research shows otherwise. Student scientists are documenting edible and medicinal plant densities far higher than in the low country, indicating that prehistoric foragers in the mountains harvested more food with a lower expenditure of energy. CWC uses ArcGIS to map those various plant communities and ecozones with precision across the landscape from valley to valley and at various elevations along the length of the Wind River Mountain Range. Student researchers map the past as adeptly as they map the present. They use GPS to record locations of 2,000-year-old white pine stumps as part of a paleoecology project that identifies changing climatic and treeline fluctuations. They will integrate this data into ArcGIS to create a tree by tree map of an ancient ghost forest.

The evident remnants of campsites and long-term dwellings called *wikiups* offer insight into high, sophisticated human population densities over time. Modern perceptions that only a few struggling human populations used the mountain wilderness do not stand up to new data being collected by CWC and other researchers. For archaeological research, accurate, objective, empirical data is critical, especially when the discoveries contradict the accepted models describing prehistoric human behavior. For example, CWC archaeologists discovered a bison jump where communities of prehistoric men, women, and children cooperated to drive herds of buffalo over an escarpment to obtain tons of food at 11,000 feet above sea level in the Dinwoody.

When humans first visited the Dinwoody Glacier, giant mammoths still roamed North America. Archaeological evidence indicates that people continued to visit and worship at the glacier throughout the Holocene. The artifacts they left behind, combined with geological data, allow researchers to track the recession of the glacier up the canyon as it melted. This map shows the retreat of the glacier toe through time.

Discovery of a series of prehistoric sites along most of the trail leading to Gannett Peak and the Dinwoody Glacier indicates that people lived and foraged for food at elevations of up to 12,500 feet above sea level in the Wind River Mountains. Projectile points chipped from stone are used to date sites through the entire span of human presence in North America to the middle 1800s at the end of the Little Ice Age.

10,000 BP

9,000 BP

6,000 BP

150 BP

BP means years "before present"

This elevation is 2,400 feet higher than the next highest documented jump, and such site complexes are universally described as a plains, not montane, hunting strategy. To convince skeptics, the team must document cultural material in 3D, and with decimeter accuracy. They must record each stacked stone cairn, all the hunting blinds and shaman pits, the locations of lithic (stone) artifacts, and organic materials of bone and wood scattered across several square miles. The next stage in this data-gathering quest will use kite aerial photography to collect high-resolution and infrared imagery of the driveline and related features.

Students apply complex geospatial technologies and methodologies under difficult conditions, and then use various ArcGIS applications to create maps and analyze data. These student researchers—coming from as far as the nation's coasts and as near as the Wind River Indian Reservation—provide data and subsequent interpretation that will rewrite textbooks. The archaeological process depends on advanced skills in geographic information science.

Buffalo jump sites like the one documented here provided ample food for several months. Maintaining this jump and reaping the benefits of each kill required the labor and cooperation of hundreds of Native Americans who spent months at a time living in nearby high-elevation villages. Currently, this site is an intriguing anomaly. But researchers could identify a significant regional subsistence strategy if they find similar sites elsewhere. But where to look? And what to look for?

Spatial analysis can dramatically change the story of how native peoples used the mountains, and guide the direction of future alpine archaeological expeditions. Using ArcGIS spatial analysis, archaeologists Burnett et al. (2009) accurately predicted the locations of lithic archaeological sites in the Absaroka Mountains on the east side of Yellowstone National Park. CWC researchers have successfully applied their model in areas of the Wind River Mountains.

A student-led modeling project using ArcGIS attempts to predict where additional buffalo jump sites might be located across the entire Greater Yellowstone Ecosystem. Because the Dinwoody site is the only known high-elevation jump site, the foundations for this predictive model stem from a small pool of geological and environmental observations. Based on that data, the model was created by isolating landforms located between elevations of 8,000 and 12,000 feet, with alpine scrub and grass vegetation, annual precipitation values of less than 35 centimeters,

and a slope of less than 20 degrees. Next, the model selected regions that have a continuous geographical connectivity and an overall area greater than 10 square kilometers (which would provide an adequate grazing area to attract herds of bison). The model also considers elk migration routes known to provide large ungulates with the easiest access corridors into and through the high country.

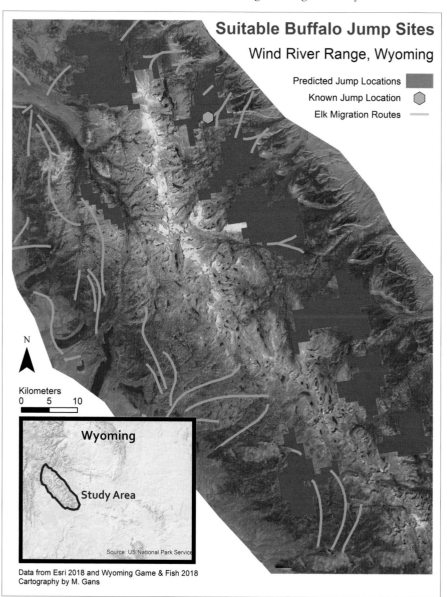

Data from Esri 2018 and Wyoming Game & Fish 2018
Cartography by M. Gans

ArcGIS spatial analytics identified all the areas in the Wind River Mountains that met the conditions for buffalo jump sites and predicted nine suitable jump areas. The next step is for field expeditions to determine whether the model is accurate. This model is being applied to other high-elevation areas in the Greater Yellowstone Ecosystem, including Yellowstone National Park. Using this map as a starting point, high-elevation archaeologists explore the predicted locations on foot to validate (ground truth) the predictive model. Their work may well add another chapter to the story of how prehistoric peoples used the high mountains.

The Dinwoody Bison Jump overlooking the Dinwoody Canyon.

ALPINE SCIENCE INSTITUTE

The remote and rugged landscape of the Dinwoody Cirque is a difficult and sometimes dangerous environment for launching a research program. Managing 15 or more students in this environment compounds the difficulty of working here, especially given that some students are conducting research at high altitudes for the first time.

CWC's Expedition Science degree program is part of a larger education and research umbrella at CWC—the Alpine Science Institute (ASI). The ASI program integrates scientific research, field-based expeditions, and leadership development by building classroom curriculum around experiential learning. The ASI embraces contemporary science and technical skills in disciplines of environmental science, geospatial information science and technology (GIST), outdoor education, and archaeology. ASI field projects allow students to travel and study in Wyoming's most remote environments.

Students in the Expedition Science program perform research in paleoecology to reconstruct past human and physical environments, and apply contemporary ecological study to the Greater Yellowstone Ecosystem. Field options include alpine glaciology, high-elevation archaeology, and the ecology and microbiology of extreme environments. In each subject area, students apply field and lab research techniques to contemporary environmental and anthropological questions. Students present their data at conferences and share it with state and federal land management agencies.

This program prepares students for careers in natural resource management, environmental science, outdoor education, glaciology, archaeology, and geospatial information science. This degree is designed for transfer to the University of Wyoming and other four-year institutions.

Packhorses hauled the team's scientific instruments, montaineering gear, and extra food most of the way, but the supplies had to be carried on foot the last three miles.

Jacki Klancher (left) and Darran Wells (right) confer at the high base camp. For the two expedition leaders, tools including GPS and accurate maps serve the campaign's top priority: participant safety.

At the required "snow school," Adam Frank, a veteran student leader, demonstrates the fine points of tying an essential Prusik knot, a climbing aid used to attach a loop of cord around a fixed rope for safety.

ENDNOTES

Adams, R. 2010. "Archaeology with Altitude: Late Prehistoric Settlement and Subsistence in the Northern Wind River Range, Wyoming." PhD diss., University of Wyoming.

Burnett, P., and L. C. Todd. 2009. *Using an Archaeological Predictive Model to Design Sample Surveys Following Forest Fires*. Poster presented at the Ninth Biennial Rocky Mountain Anthropological Conference, Gunnison, Colorado.

Chase, A., L. Kennedy, and J. Klancher. 2015. "Interdisciplinary Climate Change Expedition (ICCE): Impacts of Glacial Recession on the Chemical, Physical, and Biological Properties of the Upper Dinwoody Creek." *Wyoming Scholars Repository.*

Dahms, D. E., P.W. Birkeland, R.R. Shroba, C. D. Miller, and R. Kihl. 2010. *Latest Quaternary Glacial and Periglacial Stratigraphy, Wind River Range, Wyoming*. The Geological Society of America Digital Map and Chart Series 7.

Greenwald, A., L. Tomme, C. Schmitt, and J. Klancher. 2015. "ICCE: Black Carbon Analysis from the Dinwoody Glacier." CWC Interdisciplinary Climate Change Expedition. Published online through the University of Wyoming UGRD online repository.

Hartman, G., T. Adderhold, S. O'Sullivan, M. Baur, C. Schmitt, and J. Klancher. 2016. "Measuring Black Carbon, Water Quantity, and Water Quality in the Dinwoody." CWC Interdisciplinary Climate Change Expedition. Published online through the University of Wyoming UGRD online repository.

McGlynn, I., A. Parsekian, M. Hamrick, D. Wells, and J. Klancher. 2016. "An Analysis of Ice Depth on the Dinwoody Glacier Using Ground Penetrating Radar." CWC Interdisciplinary Climate Change Expedition. Published through the University of Wyoming UGRD online repository.

Meier, M. F. 1951. "Glaciers of the Gannett Peak." Master's thesis, Iowa State University, Department of Geology.

Stirn, M. 2014. "Modeling Site Location Patterns Amongst Late Prehistoric Villages in the Wind River Range, Wyoming." *Journal of Archaeological Science*, 41: 523–32.

Strubhar, A., A. Parsekian, A. Frank, C. Younger, M. Bostick, D. Wells, and J. Klancher. 2018. "Utilizing Ground Penetrating Radar to Conduct Subsurface Analysis of the Dinwoody Glacier and Lower Dinwoody Snowfields." Published through the University of Wyoming UGRD online repository.

Welleford, N., B. Carr, J. Klancher, D. Wells, and M. Bostick. 2015. "Exploratory Application of Ground Penetrating Radar on the Dinwoody Glacier." CWD Interdisciplinary Change Expedition. Abstract and poster published online through the University of Wyoming.

Wyoming State map designed by John Nelson, Esri.

Photo credits: Christian Harder, Benjamin Storrow, Todd Guenther, and Mara Gans.

The First Mountaineers photos courtesy of Central Wyoming College Anthropology Museum.

PANAMAPPING: GIS FOR CONSERVATION SCIENCE

Geographic information system (GIS) technology supports conservation goals in Panama by revealing how physical features of the landscape interact with current and historical human uses of the land, allowing conservation managers to visualize and communicate processes of forest change, locate critical areas, and plan conservation activities.

By Dan Klooster, David Smith, Nathan Strout, University of Redlands; Experience Mamoní; and Fundación Geoversity

The Mamoní River defines the watershed in which the Mamoní Valley Preserve works to conserve forests, streams, and the biodiversity and human communities that depend on them. Students and faculty at the University of Redlands used GIS mapping and analysis to support this goal.

MAPPING AND SPATIAL ANALYTICS FOR CONSERVATION RESEARCH

Forests of the humid tropics shelter biological and cultural diversity, protect watersheds, and sequester carbon from the atmosphere. However, they are threatened by deforestation and climate change, and so conserving them often requires strategies to protect critical areas while promoting resident human communities that are invested in conservation and sustainable uses. GIS supports these goals by revealing how physical features of the landscape interact with current and historical human uses of the land, allowing conservation managers to visualize and communicate processes of forest change, to locate critical areas, to plan conservation activities, and to place sustainable development projects.

In an ongoing series of field courses starting in May 2017, undergraduate students, faculty, and GIS professionals from the University of Redlands have been working in Panama with the Mamoní Valley Preserve (MVP), using GIS to support the reserve's goal of conserving biodiversity while promoting local culture and a sustainable local economy. Working in micro-watersheds defined by tributaries of the Mamoní River, each field course conducts parallel and overlapping research projects. These include:

global positioning system (GPS) mapping of trails and watercourses and micro-watershed boundaries; collecting original high-resolution 3D land cover data with unmanned aerial vehicles (UAVs); land cover mapping; carbon mapping; and watershed analysis.

So far, these projects have provided data, analyses, and visualizations showing reserve managers the locations of trails, streams, watershed management units, pastures and forest types, areas of similar carbon content, and the parts of watersheds that are most vulnerable to erosion. One future project will use high-resolution 3D imagery from UAVs to plan bird-watching trails. Another will do field verification of valley-wide land cover change maps using radar imagery from the European Space Agency and then use historical land cover imagery and local interviews to better understand the social history of deforestation and forest recovery. Over time, the idea is to build a thorough geographic database of increasing value for reserve management. Meanwhile, each iteration of the field course provides a rich environment in which doing GIS becomes a vehicle for undergraduate students to learn conservation science.

Mamoní Valley Preserve, Panama.

Costa Rica

CARIBBEAN

Bocas
del Toro

David

Mosquito Gulf

Gulf of Chiriquí

Coiba Is.

Colón

Panama Canal

Mamoní Valley
Preserve

Azuero

Chitré

Panama
City

Tabasará Mountains

Gulf of Panama

San Blas Range

Azuero
Peninsula

PACIFIC

Isla del Rey

La Palma

Sapo Mountains

Yaviza

Darién Mountains

Colombia

The isthmus of Panama is the narrow bridge of land that connects North and South America. Embracing the isthmus and more than 1,600 islands off its Atlantic and Pacific coasts, the tropical nation of Panama is most known for the Panama Canal, which cuts through its midsection.

THE CONSERVATION PROBLEM IN THE MAMONÍ VALLEY PRESERVE

Field courses at the University of Redlands examine a valley that is both a conservation priority and also representative of people-forest dynamics elsewhere in the humid tropics. The upper Mamoní Valley watershed covers approximately 121 square kilometers (29,900 acres) at the narrowest part of the Americas. A line drawn through the preserve from the Pacific north to the Atlantic would be only 60 km long (37 miles) and so conserving it widens the Mesoamerican Biological Corridor, a thread of land connecting North and South America. The valley acts as a buffer to the Chagres National Park and the pristine Guna Yala Comarca; it buttresses the Tumbes-Chocó-Magdalena eco-region—one of the top 20 ecological hot spots on Earth. The valley shelters the full complement of tropical forest mammals, including jaguars and four other species of wildcats, two species of sloths, several kinds of monkeys, and abundant snakes, lizards, and amphibians. Scientists from Harvard University and elsewhere have observed 290 species of birds here. The valley also appears to be a refuge for species of frogs decimated by disease elsewhere in Panama, and the Smithsonian Tropical Research Institute's Amphibian Recovery Unit studies remnant frog populations and restoration techniques.

Recent history has turned the valley into a mosaic of pasture, 15- to 30-year-old secondary forests, and stands of old-growth forest. Starting in the 1960s, small-scale farmers and cattle ranchers began clearing forest in the valley. A rough road entered in the 1980s, and deforestation pressures began to leak out of the Mamoní Valley as settlers and the hired hands of absentee landlords crossed the continental divide into indigenous Guna territory to clear fresh lands for food crops and pastures. In a well-known example of indigenous resistance, the Guna people demarcated their lands and spent many years firmly evicting settlers from homesteads in Guna territory and turning back settlers at their border with the Mamoní Valley. In the last several decades, subsistence agriculture and small-scale cattle ranching became less economically viable, and some families began to abandon the valley. This created the threat that large-scale cattle ranchers would consolidate smaller properties and convert much more of the valley to pasture.

In 2000, in coordination with the Guna people who were looking for allies to stabilize their frontier, Nathan Gray, a social entrepreneur and sustainable development activist, established a Mamoní Valley campus of Earth Train, an environmental youth leadership training organization he had founded in 1990. Earth Train recruited allies to buy strategic parcels of forest and pasture to prevent the consolidation of pasture lands into large-scale cattle operations. These actors eventually established the Mamoní Valley Preserve, a 501(c)(3) nonprofit organization using the tools of landownership, conservation easements, and landowner pledges to conserve the extent and quality of the valley's forests and waters, to strengthen biodiversity, to improve the health of the forest's inhabitants, and to promote a sustainable local economy. Currently, the reserve has the explicit support of the landowners in possession of about half the valley. The goal, however, is to extend forest conservation and sustainable development to the entire upper Mamoní River watershed.

RESEARCH PROJECTS

Through an overlapping set of parallel projects, the University of Redlands team has been gathering original data with GPS, UAVs, and forest measurements, and analyzing their results in the context of previously available geographic data to produce maps of land cover, watersheds, and carbon content. "Where?" questions guide these projects. Where are trails and stream courses and the boundaries between micro-watersheds? Where are pastures and different types of forest? Where are younger, older, and critical types of forest? Considering the amount of carbon in each land cover type, how much carbon is in each micro-watershed and where is it located? Considering the land cover and terrain within each micro-watershed, where are the places most vulnerable to erosion? To answer these questions, students use GPS, UAVs, and forest surveying techniques to generate original data, and then they interpret and combine this data to create new ways of representing the landscape.

GPS: From footprint to map

Using GPS, students turn walking into a mapping activity. Upon arriving in the Mamoní Valley, they quickly realized the need for more detailed location data on the trails, watercourses, ridgelines, and watershed boundaries of the area. These are crucial forms of data for understanding the Mamoní Valley Preserve landscape. Students carried hand-held Garmin 60 and 62 GPS units on all their travels over the landscape, capturing both linear features and points of interest such as significant land cover changes and man-made infrastructure. GPS was used to also determine the location of their research plots where forest mensuration measurements were recorded in field notebooks.

Upon returning to the research station, students exported their GPS data to shapefiles using DNRGPS (an open source application) and imported them into ArcGIS Pro for editing. After reviewing the data, students became aware of the limitations of the GPS units, especially under heavy tree canopy. A discussion of GPS accuracy and precision ensued with a decision to smooth out the trails and watercourses to account for the lack of accuracy in areas of dense forest. Nevertheless, original GPS data proved useful for improving accuracy of existing maps, which often misplaced minor streams and included no micro-watershed boundaries. This mapping activity also added several important trails to Mamoní Valley Preserve maps.

University of Redlands student-researchers used GPS to record the location of forest measurement plots. In each plot, they counted trees, measured tree height and diameter, and used this data to calculate carbon content per hectare. GPS locations later allowed them to correlate their field measurements with land cover maps.

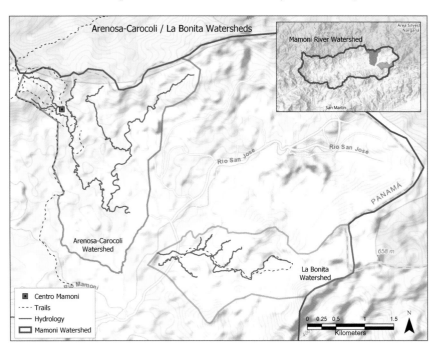

In addition to the trails and watercourses collected by the students, two micro-watersheds, the Arenosa-Carocoli and the La Bonita, have been delineated. These two watersheds formed the basis for the research in the valley.

Student-researchers worked in groups to transfer field data from GPS units and notebooks to spreadsheets, and then to a GIS.

UAVs: A high-resolution view from above

To truly appreciate the complexity and diversity of the rain forest, you must immerse yourself in it and experience it—but to understand the structure and health of the rain forest, you need to get above the tree canopy and see it on a much broader scale. Thanks to new and innovative UAV, or drone, technologies, this can now be done quickly, easily, and inexpensively. Within just a few years, UAVs have revolutionized GIS data collection—providing near-real-time imagery products at very high resolutions. Using these imagery products, the class can evaluate forest types, density, and maturity; identify and assess areas of deforestation; monitor forest restoration efforts; and much more. Drone image acquisition also allows us to get below the cloud cover that so frequently interferes with standard satellite platforms, especially in the humid tropics.

An added benefit to using drones in GIS instruction is that students have the opportunity to be involved in all aspects of image acquisition and photogrammetric processing of the imagery that they ultimately use in their analyses. This gives them an understanding and appreciation of the processes and assumptions involved in the creation of the imagery products that they interact with on a daily basis.

Over the course of five days of image acquisition in 2018, the team dealt with several challenges, including access to a central takeoff/landing location, constantly changing weather conditions, malfunctioning equipment, and difficult terrain to pilot over, but the team managed to cover the majority of the 3 km² La Bonita micro-watershed. This processed imagery expanded imagery collected in the neighboring Arenosa-Caracoli micro-watershed and surrounding areas in 2017.

Two DJI Phantom 4 Professional UAVs were used to acquire roughly 4,500 photos above the study area. These quadcopters are easy to fly and they take off/land vertically—very important features when flying in dense rain forest with students. The Phantom 4 Professionals have a 1-inch 20 megapixel sensor for a high-resolution product and also a mechanical shutter to eliminate the distortion often introduced flying at higher speeds with a rolling shutter. The UAVs handled themselves remarkably well given the conditions, and the group was satisfied with the field performance of these vehicles.

Flight planning was done using Pix4D Capture on an iPad at the headquarters in Centro Mamoní with internet connectivity, but changes were regularly made in the field due to changing conditions. Consumer flight planning apps such as Pix4D Capture require that missions fly at a constant altitude above the takeoff point, set by the operator. The image resolution and overlap calculations are made based on this altitude and assumed constant because the UAV isn't aware of the changing ground elevation underneath it. This presents a challenge when mapping a large area with varying terrain because the elevation of the ground under the UAV is changing and therefore the image overlap and resolution is difficult to plan for. To resolve this issue, missions were planned in "terraces" following topo contour intervals. This approach helped reduce the effects of the overlap and resolution issues but also significantly increased the number of missions, the total flight time, and general complexity of the image acquisition project.

Pasture areas made ideal launch sites for DJI Phantom 4 Professional UAVs. These drones captured high-resolution imagery useful for mapping pastures and different types of forest.

Back at Centro Mamoní, the team downloaded the images to a hard drive and began processing the missions using Pix4Dmapper. Pix4Dmapper is a photogrammetry software for generating and analyzing orthomosaics and other mapping products from images collected via UAV. The orthomosaics for this project vary in resolution based on flight altitude and the changing terrain, but can all be categorized as submeter resolution. The software also generates 3D products such as point clouds, meshes, and DTM output that can be used for further analyses such as tree canopy heights.

Pix4D provides both desktop and cloud solutions but cloud-based tools were impractical given limited internet bandwidth and sporadic downtime. Desktop processing was also difficult due to the limited computing resources and several power outages. Mission processing would often take multiple days/nights to complete, so the team had to be strategic to prioritize areas to fly and process.

The orthomosaics and other imagery products were not completed and readily available for tasks such as land cover classification, but the team did their best to create draft products. Another approach was to use ArcGIS Desktop's GeoTagged Photos to Points tool to generate a point layer with a point for each photo linked to the full-resolution photo in a pop-up window in ArcGIS Pro. Students ultimately used a combination of Pix4D-generated orthomosaics and the photo points layer with linked photos to complete tasks such as land cover classification.

Once processing was completed, the orthomosaics were imported into an ArcGIS mosaic dataset with time stamps. This dataset will continue to be used and extended in the future to display the datasets in a single mosaic while preserving overlaps as well as query and filter over time to support operations such as change detection. This dataset now covers roughly 9.25 square kilometers of high-resolution aerial imagery and can be used for further research and land management.

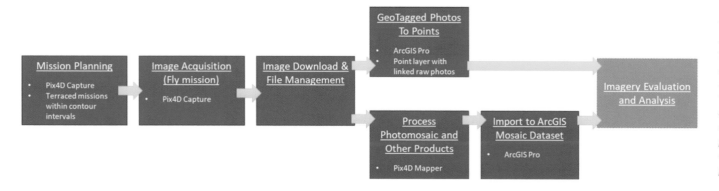

The workflow for UAV imagery capture and processing required software for flight planning, imagery capture, and the imagery processing into orthomosaics. Limited computer processing ability at the Centro Mamoní rain forest research station delayed imagery processing, however. GeoTagged UAV photographs provided a work-around to interpret land cover where orthomosaics were incomplete.

UAVs captured highly detailed land cover data, such as this sample of processed imagery from within the La Bonita micro-watershed.

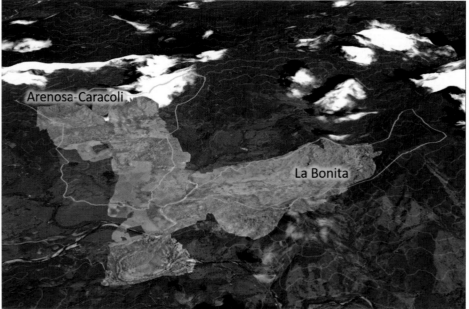

During the first two years of the University of Redlands field courses, student-researchers collected about 4,500 images from quadcopter UAVs and used photogrammetry software to create a geographically corrected, high-resolution orthomosaic covering more than 9 square kilometers. The data now provides detailed landcover information on two important micro-watersheds contributing to the upper Mamoní River.

Land cover and carbon mapping

The orthomosaics generated by UAV remote sensing provide highly detailed visual information that is intrinsically useful even in its raw form. However, GIS visualizations and analyses also benefit from an analytic interpretation of land cover that identifies areas of generally similar characteristics.

To produce such a land cover map, researchers worked with a professional forester familiar with the area. They learned to understand and recognize forest stands and chronosequences. A forest stand is a community of trees generally similar in composition and age, while a chronosequence is the series of successional phases a forest stand goes through as it recovers from a disturbance such as being cleared for agriculture. Students learned to recognize key land covers in the field and on the raw UAV orthomosaic, and then used GIS to delineate land covers and produce a land cover map.

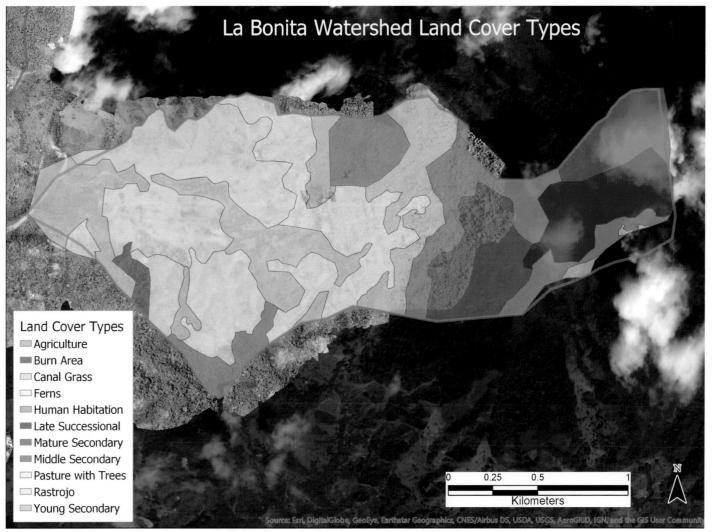

Land Cover Types
- Agriculture
- Burn Area
- Canal Grass
- Ferns
- Human Habitation
- Late Successional
- Mature Secondary
- Middle Secondary
- Pasture with Trees
- Rastrojo
- Young Secondary

Students at the University of Redlands produced land cover maps showing a sequence of land cover types found in La Bonita, ranging from pasture through different ages of forest regrowth, including areas of very old, late successional forest. Resulting maps reveal a landscape shaped by a history of deforestation for pasture in valley bottoms and forest conservation in the more remote headwaters of streams such as La Bonita.

Carbon sequestration and forestry

Carbon emissions from deforestation are a major contributor to global greenhouse gas emissions, while growing forests sequester carbon, and so conserving forests provides an important environmental service. Mamoní Valley Preserve managers wish to visualize and communicate how much carbon they are managing and where it is located. Carbon maps could also tell them where to concentrate conservation efforts to avoid carbon emissions from deforestation, and where they might best invest in reforestation projects that could sequester carbon from the atmosphere. To address this question, student-researchers first worked with a professional forester to learn essential forest measurement skills. They next formed teams and gathered data from sample plots, including the height and diameter of all trees within the plot. They entered the data into a spreadsheet and calculated the volume and carbon content of the trees in each fixed area plot. The forester selected plots to represent the various land covers.

Students used GPS data to confirm the correspondence of sample plots with their land cover map and then took the average of the results from the plots in each land cover type. They then assigned the average carbon content value to each land cover. Finally, they joined the carbon value data with the land cover data to produce a new map representing the distribution of carbon within each watershed studied.

The largest trees in the Mamoní Valley are found in late successional forests. These forests also contain the most carbon per hectare.

Carbon mapping integrated multiple research projects. Land cover mapping with UAV (drone) imagery produced a map layer showing pastures and different-age forests. Meanwhile, student-researchers measured carbon content of forest plots representing different land covers. Joining the sample carbon data with land cover produced the carbon map.

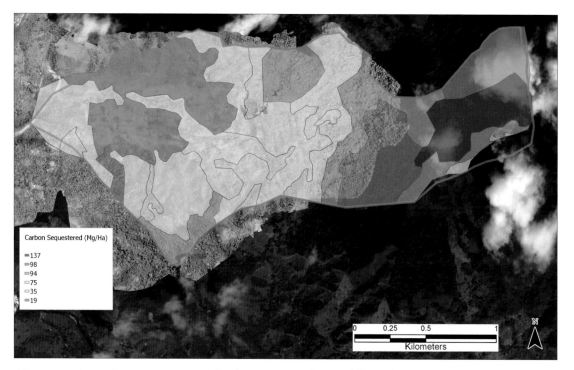

Carbon Sequestered (Mg/Ha)

- 137
- 98
- 94
- 75
- 35
- 19

After measuring carbon content in sample plots corresponding to different land covers, student-researchers produced carbon maps for each micro-watershed they studied. Their results indicate the highest carbon density is found in the oldest forests in the headwaters of micro-watersheds, and the least carbon is in the pastures.

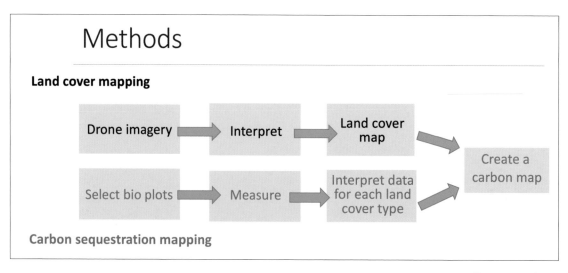

Methods

Land cover mapping

Drone imagery → Interpret → Land cover map →

Select bio plots → Measure → Interpret data for each land cover type → Create a carbon map

Carbon sequestration mapping

Watershed analysis

Efforts to conserve and improve the forests and waters in the upper Mamoní Valley watershed require an understanding of the watershed dynamics at the micro-watershed level. As such, student-researchers analyzed the erosion vulnerability of each micro-watershed studied and answered six questions for each target watershed:

- Where are the micro-watershed boundaries?
- Where is the main stream and where are the significant tributaries?
- Where do streams have riparian forest buffers and where are these absent?
- Where are the forests and pastures in this watershed?
- Where are steep slopes and flat areas?
- Where are steep slopes covered in pasture and young forest regrowth, and thus most vulnerable to erosion?

GIS provides essential tools for answering these questions. The first step is delineating the micro-watershed. Students analyzed a digital elevation model (DEM) to identify initial watershed boundaries, and in areas of abrupt topography, this method was sufficient. Where topography was flatter, however, they needed to hike watershed divides using GPS to trace the high points separating one watershed from another. Existing data on streams contained numerous errors, so students also hiked the creek beds with GPS to create more accurate data on stream locations.

Second, students mapped riparian barriers by performing visual analysis of the UAV orthomosaic. During exercises mapping streams in the watershed with GPS, students occasionally stopped in the middle of a stream to take measurements with a densiometer, an instrument that is used to estimate canopy cover, recording the location with a GPS. Those geo-located observations allowed them to verify their interpretations of riparian buffers from the orthomosaic covering the same area.

The third major component of the watershed project was to visualize and represent the topographic shape of the watershed. Using the DEM, students created vertical profiles showing the descent of the major stream and the shape of cross sections at various points along the valley. In addition, student researchers also used the DEM to generate a map showing the slope of each hillside and mountainside in the valley.

Fourth, students combined the slope map with land cover. The previously created land cover map showed the locations of pastures and different ages of forest. By combining the two maps, students were able to locate areas of special concern, such as steep slopes under pasture or early forest regeneration.

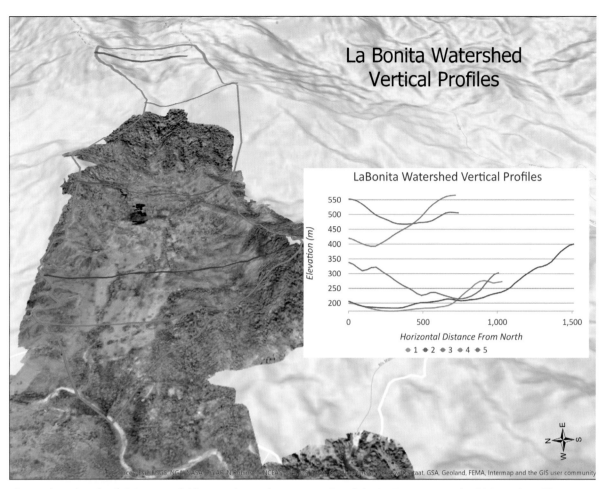

The analytical tools of GIS are useful for visualizing the topography of the La Bonita watershed. Vertical profiles represent the valley slope along illustrative cross sections.

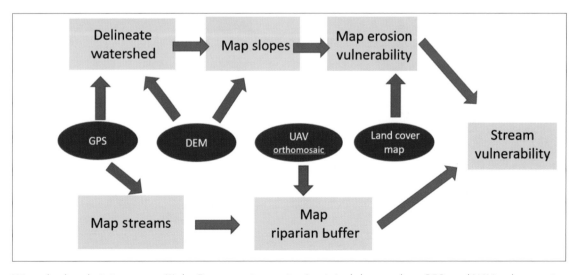

Watershed analysis integrates (1) the Panamapping project's original data, such as GPS, and UAV orthomosaics; (2) maps made with the data, such as stream courses and land cover maps; and (3) analysis of slopes using DEMs. By comparing slope data and land cover data, student researchers could identify vulnerable land covers on steep slopes. Sections of stream without riparian buffers immediately below vulnerable slopes would be most at risk from erosion.

Results

Each of these projects helps MVP managers meet their goal to conserve the extent and quality of the valley's forests and waters. GPS tracking, GIS maps, and UAV remote sensing provide more detailed and accurate information on trails, streams, and—especially—forest cover. The land cover map helps them more easily visualize the locations of target land cover types such as old growth forest, to correlate land covers with other data such as elevation and slope, and—by repeating the process with land cover data from different time periods—to quantify and visualize change over time. Carbon mapping reveals that the micro-watersheds so far studied store significant amounts of carbon; that most of this carbon is in relatively remote, higher elevation areas spared deforestation during the twentieth century; that there are many areas where young growing forests are capturing carbon; and that many pasture areas are potentially available for carbon-capturing reforestation projects.

Micro-watershed mapping locates essential management units in which managers can now develop, compare, and monitor useful new indices such as the percentage of a micro-watershed covered by pasture or forest. After each field course, students have presented their visualizations and analyses to MVP reserve managers.

Before the University of Redlands began its field courses, the reserve had only general maps of the valley's land cover, topography, hydrology, major roads, and property boundaries. Available geographic information existed as individual maps and often relied on data with obvious errors and frequent omissions. Data from each research project is being compiled into an integrated database of geographic information so that MVP managers, University of Redlands researchers, and others can better address conservation science and management questions for the reserve.

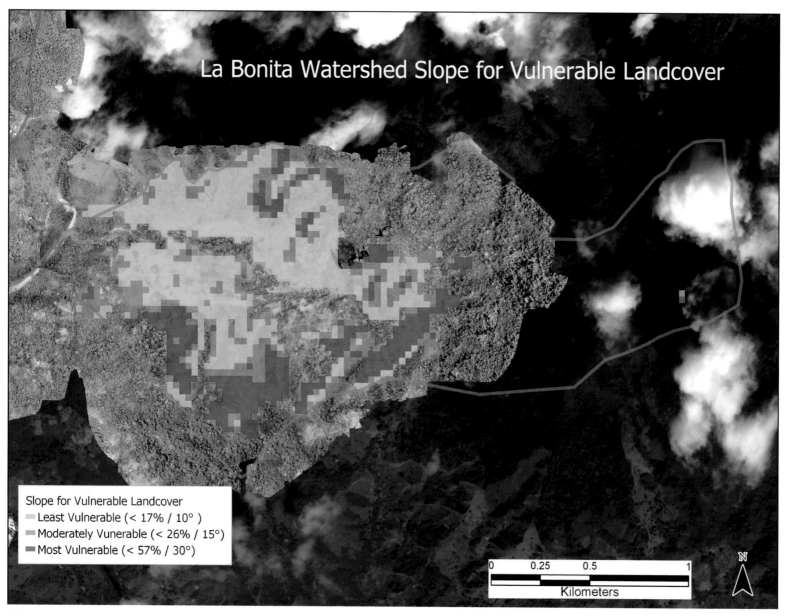

La Bonita Watershed Slope for Vulnerable Landcover

Slope for Vulnerable Landcover
- Least Vulnerable (< 17% / 10°)
- Moderately Vunerable (< 26% / 15°)
- Most Vulnerable (< 57% / 30°)

0   0.25   0.5   1
Kilometers

N

What areas in La Bonita micro-watershed are most vulnerable to erosion? To answer this question, University of Redlands student-researchers produced this map showing where pastures and recently abandoned pastures are found on very steep slopes. These areas could be the focus of reforestation efforts.

GIS IN THE STRUGGLE FOR CONSERVATION: SUPPORTING THE MAJÉ EMBERÁ

It is dark and there is no electricity. Students crowd around a makeshift table under a palm-thatched roof. Only our headlamps illuminate the map we are using to plan tomorrow's UAV flight. Dressed in his traditional garb, Cacique (Chief) Lorenzo orients himself on the map and points out the village site to University of Redlands UAV pilots. He shows us where ancestral lands were flooded out in 1976 to provide hydroelectric power to Panama City. "We had to give up our lands so the rest of the country could have electricity, but we don't have electricity, or water, or even rights to the remnant of our territory that we live on," Lorenzo tells his visitors. With his finger, Cacique Lorenzo traces the watershed boundary on a map. It shows the GPS track of the hikes he and his brother took along the watershed boundary, revealing significant errors in the official boundary of the Majé River conservation area.

Lorenzo points out areas where his people used to collect medicinal plants to support pregnant women and to cure the sick. Many of those areas are deforested now. The problem, he explains, is that his people's remaining lands have been declared a conservation area, and this means they cannot officially own them, protect them, or manage them. But conservation authorities don't protect the lands from ranchers and loggers. Settlers have cleared large areas of pasture near the village, and a fire set to clear forests for pasture spread into the valley of the creek the Majé Emberá use for drinking water, destroying their aqueduct.

Cacique Lorenzo's map was produced by the Danish NGO Forests of the World, and it shows areas of land cover change calculated using radar remote sensing from a European Space Agency satellite. Visitors from the University of Redlands are there to support the Majé Emberá by gathering data to help verify and illustrate the larger image. Lorenzo asks them to use their UAV to document a recently deforested area and gather UAV data showing a forest he knows will soon disappear. The Majé Emberá will use the UAV imagery to support their ongoing campaign to defend their territory from encroachment and to help them lobby the Panamanian government for land rights to defend what is left of their ancestral territory. University of Redlands students, meanwhile, are seeing the role GIS can play helping indigenous people engaged in conservation struggles.

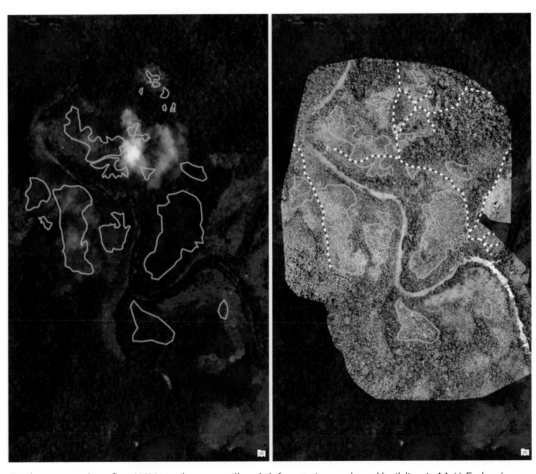

Student-researchers flew UAVs to document illegal deforestation and road building in Majé Emberá lands. The image to the left shows 2016 imagery from ArcGIS Online, with polygons showing the forest cover lost by May 2018. The right image shows UAV images from May 2018, recently deforested areas and illicit roadbuilding.

With the direction of indigenous leaders who orient student-researchers to their needs, GIS and UAVs can document processes of deforestation that threaten both conservation and the indigenous rights to land and livelihood.

GIS FOR CONSERVATION EDUCATION

Through these projects, students are building an enterprise GIS for the Mamoní Valley Preserve, including land cover, watershed analysis, landownership, carbon storage, hydrology, and trails. Their use of GIS tools contributes to conserving the rain forest, but at the same time it also advances students' education in conservation science. Each iteration of the course becomes a vehicle for intensive learning through which students use GIS to build an integrated understanding of conservation challenges and techniques, based on direct observations, field training, remote sensing data from UAVs and satellites, original field measurements, making maps, information from lectures, and constant discussion with course instructors and peers. By doing GIS, students learn important aspects of conservation science.

GPS mapping gets students into the forest, walking along trails, tracing ridgelines, and wading in creeks. It teaches an important data entry skill for GIS and promotes concrete appreciation for accuracy and precision—students become savvier, appropriately critical users of spatial information. Flying UAVs also has clear educational value. It gives students additional experience integrating original data into a GIS while introducing optical remote sensing and a compelling new geospatial technology.

Using GIS to analyze watersheds, students think about the connections in a landscape. They visualize links between land cover, slope, and erosion. They consider the role of riparian barriers in buffering streams from the runoff from cattle pastures. Using GIS tools, they can better visualize, interpret, and communicate the watershed: Here is an area where a steep slope has been deforested! Here is where lack of riparian buffer makes the stream vulnerable to erosion and nutrient enrichment from that pasture!

Using GIS to map forest carbon provides additional learning opportunities for undergraduate students. Tasked with answering the question "Where is the carbon?" students work backward, first developing an understanding of the science of forest succession in a temporal, nonspatial sense, then developing basic skills in the science of forest mensuration. Next, students combine field measurements of forest density with their land cover map showing units of forests of similar type and age. This allows landscape-level calculations of carbon content and landscape maps of carbon densities. The process of carbon mapping reinforces important concepts of carbon sequestration and the role of forests in global climate change mitigation strategy. But perhaps more importantly, it immerses students in the scientific process of data gathering as they learn to estimate the heights of trees, measure their diameters, count individual trees in a given area, and defensibly extrapolate their findings to a land cover type and then to a land cover map. Students also learn to recognize the characteristics of young stands of trees compared with old ones so they begin to perceive the role of forest succession in creating a landscape. As they create and interpret land cover maps, they learn about the social process of deforestation, conservation, and forest recovery. In Panamapping, GIS becomes a powerful vehicle of scientific education.

GIS and UAVs capture data and permit analysis useful for conservation science. They also create opportunities for collaboration and learning between students and indigenous people.

Acknowledgments

Panamapping courses would have been impossible without the irreplaceable contributions from the student researchers whose work is presented here: Michael Adami, Zachary Booker, Harrison Cannon, Serena Dudas, Taiga Gamell, Kyle Garrity, Javier Gasga, Nicholas Graff, Enrique Huerta, Chris Merritt, Sarah O'Connor, Torrey Rotellini, Zachariah Smith, Meera Srinath, Will Tyrell, Graham Young, Akio Anderson, Luis Salgado, Joe Imburguia, Kaitlyn Dreissnack, Jessica Pierce, Jordan Hooey, Veronica Creed, Abigail Bohman, Ian Matsumoto, Sumner Macpherson, Taryn Fowlkes, Ava Gotthard, and Sarah Graham. Our safety, comfort, and success in the field depended on the work of Mark Knetsch, Ana Gili, Nico Armstrong, Ezequiel Coniglio, and Gabriel Salazar. Kike Arias, Matthew Brewer, Venicio Wilson, and Tim Krantz provided crucial instructional enhancements. Lazaro Mecha, Narciso Mecha, and Clemente Mecha hosted student researchers in their village. Finally, Steve Moore and the University of Redlands Center for Spatial Studies provided crucial equipment and GIS support.

PART 5
TECHNOLOGY SHOWCASE

This book has already shown how science goes hand in hand with technology (engineering). One of the most exciting trends of the modern age is how science leverages the exponential power and assistance of artificial intelligence (AI) to help address the unprecedented challenges facing humanity and the planet, including climate change, water scarcity, global health crises, food security, and loss of biodiversity. GIS technology is no different; it extends our minds by abstracting our world into knowledge objects that we can create, replicate, and maintain. These knowledge objects include data, imagery, and models that explain process and workflows, as well as maps that communicate and persist in apps. Enjoy this section of vignettes on GIS technologies that help create new systematic frameworks for scientific understanding.

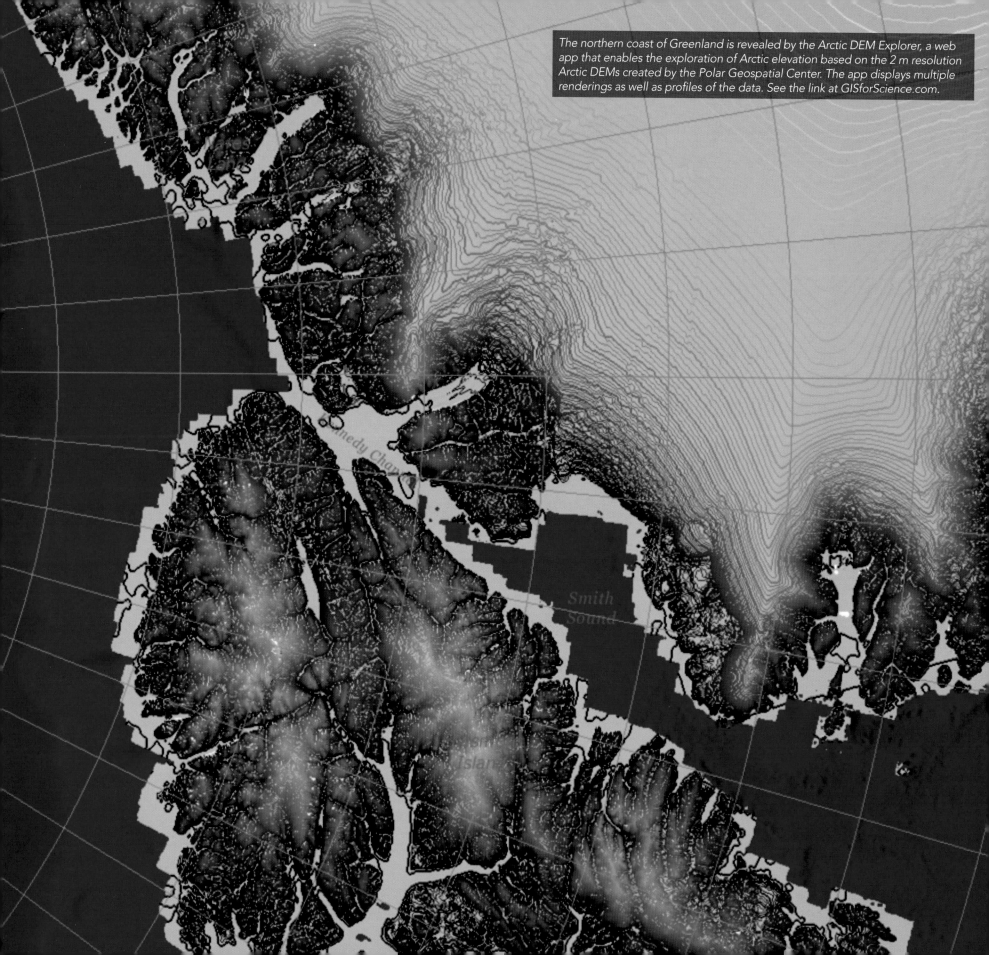

The northern coast of Greenland is revealed by the Arctic DEM Explorer, a web app that enables the exploration of Arctic elevation based on the 2 m resolution Arctic DEMs created by the Polar Geospatial Center. The app displays multiple renderings as well as profiles of the data. See the link at GISforScience.com.

Kennedy Channel

Smith Sound

Island

EMERGENCE OF THE GEOSPATIAL CLOUD

A geospatial cloud for science

The combined forces of the internet and cloud computing are clearl impacting—some would say *transforming*—how scientists and researcher do their work. With the geospatial cloud, they can easily and inexpensivel access an immense amount of geographic information on almost any subjec and take advantage of cloud computing resources to perform analysis and mapping. They can also easily share and combine their data with other shared data in lightweight apps to build community and collaboration.

The advantages of the geospatial cloud include scalable computing, storage of large datasets, big-data computation, and the ability to gather resource during critical events, such as disaster response. The geospatial cloud give scientists and researchers the ability to handle big analytical problems and larger datasets from multiple sources more efficiently. Scientists also can analyze real-time inputs from an increasing array of sensors.

Big real-time data

Most of today's real-time and big-data problems are inherently geospatial Examples include Internet of Things (IoT) sensors; devices such as smartphones vehicle sensors; cameras; and imaging sensors on drones, airplanes, and satellites.

Real-time data streams from sensors provide instant context and help users display information dynamically as spatial dashboards. Spatial information can trigger real-time responses. While the computing requirements for situationa awareness can be demanding, the geospatial cloud can easily store billions of records that users can query to ask complex questions and perform location analytics.

A geospatial gateway to artificial intelligence

The geospatial cloud opens a gateway to emerging technologies, such as artificial intelligence and machine learning. These technologies help us understand and solve problems related to imagery classification, spatial pattern detection, and predictive modeling. Data scientists use AI for analytic and visualization to integrate science into a geospatial framework.

Application programming interfaces (APIs) in the geospatial cloud allow scientists to develop apps that blend their own methods and algorithms with geospatial processes. These processes help scientists perform deep analyses and create interactive visualizations in 2D and 3D. Scientists can access apps and maps from wherever they log in. And organizations can set defined roles in configured apps to ensure that the right people are working together within a secure but highly collaborative environment.

EQUAL EARTH PROJECTION

Beginning in 2017, a small team decided to create a new map projection that would represent Earth in a less distorted but visually pleasing and balanced way. It is called the Equal Earth projection and is available to everyone at equal-earth.com.

There are many types of projections, each designed to serve specific purposes. Most try to reliably preserve cartographic properties, such as direction, shape, area, and distance, so that the maps can better serve the people who use them. However, preserving one of these properties typically distorts the other properties. For example, a projection designed to preserve area measurements generally distorts the shape of some map features, such as continents.

The aesthetic quality of a map is an important consideration. When someone looks at a map, they automatically assess whether it looks right or meets with their own mental map of the world. If the country or continent they live in looks elongated or squashed, or is larger or smaller than a comparable area, their trust in the map diminishes. The team researched many existing and popular projections and narrowed down the list to those that best showed the right balance between distortion and aesthetics that they were looking for in their new projection.

The Equal Earth projection is an equal-area projection based on the pseudocylindrical model. Pseudocylindrical projections mimic Earth's spherical form with arcing lateral meridians (also known as *longitudes*). Some pseudocylindrical equal-area projections show the North and South Poles as points. While Antarctica retains its circular shape, the extreme compression in high *latitudes* where meridians converge means there is not enough room to place labels. Other projections have short pole lines instead of pole points, but map readers find the outward bulge of the lateral meridians of these projections displeasing. These and other projections came close to meeting the team's design preferences but still had aesthetic issues. For example, the Gall-Peters projection has straight parallels (latitudes) and straight meridians but greatly deforms certain map features in the tropical and polar regions.

These distortions are significantly reduced in the Equal Earth projection. The meridians are equally spaced along every straight parallel line of latitude. Lateral meridians do not excessively bulge outward, and they loosely approximate arcs that mimic the overall appearance of a globe. The projection shows the continental outlines in a visually pleasing and balanced way. The landmasses in the tropical and mid-latitude areas are less elongated, and areas along the poles are less flattened and compressed.

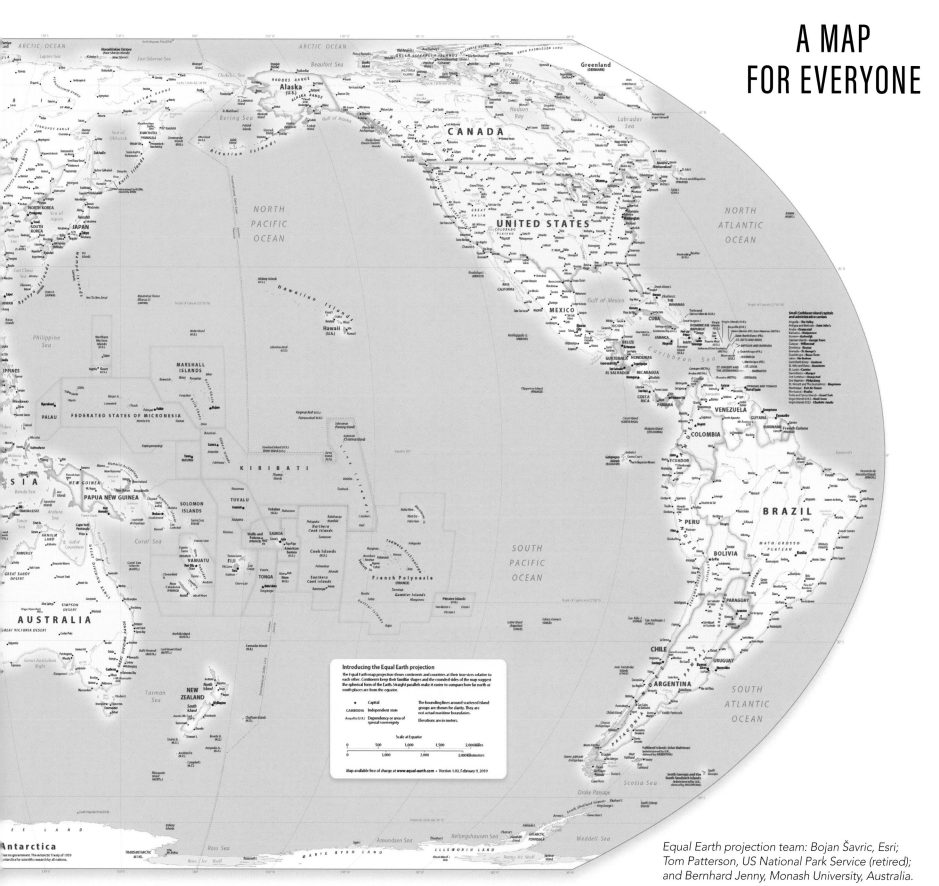

Introducing the Equal Earth projection

The Equal Earth map projection shows continents and countries at their true sizes relative to each other. Continents keep their familiar shapes and the rounded sides of the map suggest the spherical form of the Earth. Straight parallels make it easier to compare how far north or south places are from the equator.

●   Capital

CAMBODIA   Independent state

Anguilla (U.K.)   Dependency or area of special sovereignty

The bounding lines around scattered island groups are shown for clarity. They are not actual maritime boundaries.

Elevations are in meters.

Scale at Equator

0   500   1,000   1,500   2,000 Miles

0   1,000   2,000   3,000 Kilometers

Map available free of charge at www.equal-earth.com · Version 1.03, February 9, 2019

Equal Earth projection team: Bojan Šavric, Esri;
Tom Patterson, US National Park Service (retired);
and Bernhard Jenny, Monash University, Australia.

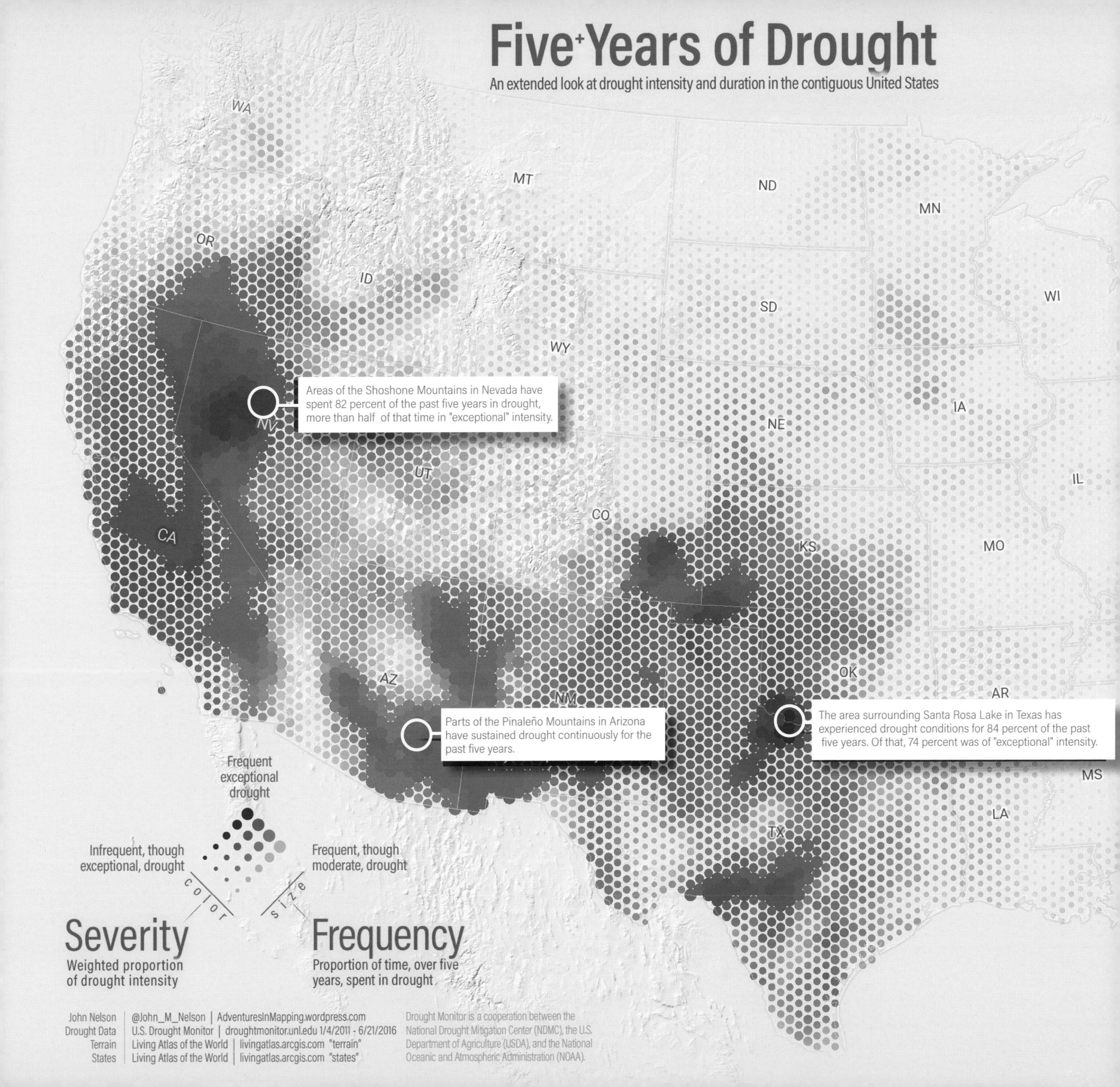

Five+Years of Drought
An extended look at drought intensity and duration in the contiguous United States
WA
MT
ND
MN
OR
ID
SD
WI
WY
NV
Areas of the Shoshone Mountains in Nevada have spent 82 percent of the past five years in drought, more than half of that time in "exceptional" intensity.
IA
NE
UT
IL
CO
CA
KS
MO
OK
AZ
AR
NM
Parts of the Pinaleño Mountains in Arizona have sustained drought continuously for the past five years.
The area surrounding Santa Rosa Lake in Texas has experienced drought conditions for 84 percent of the past five years. Of that, 74 percent was of "exceptional" intensity.
MS
LA
TX
Frequent exceptional drought
Infrequent, though exceptional, drought
Frequent, though moderate, drought
color
size
Severity
Weighted proportion of drought intensity
Frequency
Proportion of time, over five years, spent in drought
John Nelson | @John_M_Nelson | AdventuresInMapping.wordpress.com
Drought Data | U.S. Drought Monitor | droughtmonitor.unl.edu 1/4/2011 - 6/21/2016
Terrain | Living Atlas of the World | livingatlas.arcgis.com "terrain"
States | Living Atlas of the World | livingatlas.arcgis.com "states"
Drought Monitor is a cooperation between the National Drought Mitigation Center (NDMC), the U.S. Department of Agriculture (USDA), and the National Oceanic and Atmospheric Administration (NOAA).

SCIENCE OF THE HEX

Tessellation

Often, an archaeologist or crime scene investigator's first task is to create a grid over the study area, with each cell receiving its own name and known location within a neatly compact perimeter. Only then can excavations and investigations become procedural, orderly tasks for many small and manageable constituent locations. Tessellation is the GIS scientist's digital version of this process. Also known as "tiling the plane," a tessellation is a gridded mesh covering a geographic area with a consistent shape, leaving no gaps—like the tiles of a bathroom floor.

Aggregation

A tessellation is a helpful way to aggregate a geographic phenomenon into simple, bite-sized areas. A spatial join of a dataset into the underlying cells of a tessellation can create convenient local zones wherein scientists make discrete local statements.

This map uses aggregation to make big data more visually discernable by rolling hundreds of thousands of drought polygons into discrete hexagonal cells that serve as buckets for local statistical summaries and allow for cartographic visualization.

Hexagons

When aggregating a geographic phenomenon into a tessellation, hexagons are an excellent choice. The hexagon is at once a geographically compact and geometrically complex shape. Because a tessellation of hexagons creates a staggered pattern, like a wall of bricks where each row is offset by half a brick width, they are not as prone to vertical or horizontal stripe effects that can be visually distracting in tessellations with square cells. Hexagons efficiently segment the patterns of the natural world, like the combs of bees or cooling magma columns, and have an inherently crisp and appealing visual aesthetic.

Bivariate symbology

This map shows a ruinous five-year drought that afflicted the American West in recent years. A single, succinct visualization of drought intensity and duration helps scientists understand a large phenomenon more locally, and effectively communicate information to specialists and nonspecialists.

The source data for the map took the form of hundreds of thousands of drought polygons, each with its own drought-intensity attribute, making it difficult to visualize. By transforming this data into a single grid of hexagons and assigning visual symbology, the phenomenon is more likely to reach an audience with its distilled and impactful message.

Not all droughts are equal. An area may be subject to infrequent exceptional drought, frequent and moderate drought, or anything in between. How might the nuance of two, equally important, dimensions of a phenomenon be shown in a single map? The symbology in this thematic map uses color to denote accumulated drought severity, and size to denote drought frequency. Encoding multiple attributes into a single symbol is called *multivariate* symbology, or in this case, *bivariate* symbology, because two attributes are shown.

MODELING THE FOOTPRINT OF HUMAN SETTLEMENT

Persons / square kilometer

30,000

1

POPULATION DENSITY OF CENTRAL JAPAN, 2016

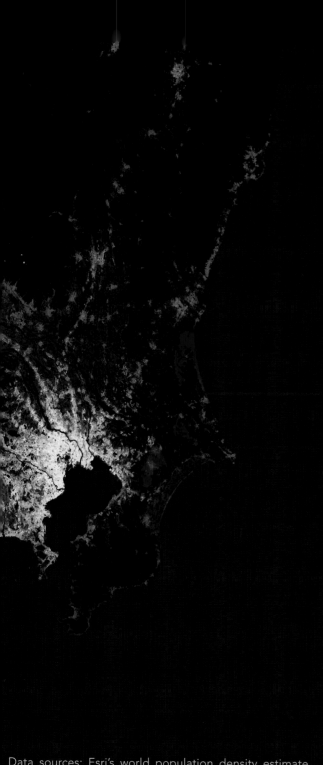

Nearly everyone lives somewhere, and each of us knows where specific people live. It stands to reason that we can map where everybody lives. The map of Japan on these pages is an excerpt from a global map that is beginning to prove this reasoning. National governments take censuses. To support that work and many other governmental functions, they maintain lists of addresses where people live. Thus, it's easy to presume that making this map would be a relatively simple task of compiling data. Yet, it is not. Unfortunately, many countries do not make their most detailed census data or lists of addresses available, so how did this map come to exist?

The short version of the story is that geographic information science, coupled with big data analysis, produced a model of the footprint of human settlement. This model is an excellent surrogate for detailed census data and addresses.

Making the map

We still use census data to make maps, though it is less detailed and less current. Because of funding issues, 55 percent of all countries do not conduct censuses where people live. Even so, such data can be proportionately redistributed onto a footprint of human settlement.

That footprint is produced using Landsat 8 imagery and a geoprocessing model that identifies and scores the locations most likely to be human settlement. The key to using Landsat 8 imagery, specifically the 15-meter cell size panchromatic (grayscale) band, comes from a surprising source: ecological modeling. Ecologists know that the roughness of terrain is an important characteristic for defining the habitats of plant and animal species. Rugosity algorithms are applied to elevation models of Earth to describe how rough the terrain is in a given locale.

Applying a rugosity algorithm to Landsat 8 panchromatic imagery indicates where buildings are because their shadows are much darker than adjacent cells. Applying a rugosity algorithm (to determine surface roughness) is a big step forward. Previous work to estimate human settlement with imagery used convolution algorithms that rotate sections of imagery to count pairs of cells. The rugosity algorithm proved to be much faster and could guarantee identifying five-by-five cell areas with at least two pairs of adjacent cells.

Processing global Landsat 8 data

Using Landsat 8 means using 8,600 specifically chosen images covering Earth's land area. Each of those images has about 250 million cells, meaning the rugosity model analyzes just over two trillion cells. The Landsat 8 program captures images in the same locations every 16 days, meaning each year the process for calculating the footprint of settlement evaluates up to 22 candidate images for each of the 8,600 locations. The process looks for the lowest amount of clouds, leaf-off conditions, and no snow.

One of the big advantages of cloud computing is that instead of downloading all these images to process them (that's more than 4.5 TB of data), ArcGIS software can reference them in a mosaic dataset and perform the analysis without downloading them. Another advantage of using mosaic datasets in ArcGIS is assigning raster functions to each dataset that automatically pre-process the cell values. In this case, the process computed a "top of atmosphere correction," which is an image scientist's way of saying, "Clean up the differences between adjacent images," which then allowed one rugosity algorithm to be applied to all images, rather than having to customize the algorithm for each image.

Data sources: Esri's world population density estimate for 2016, derived from the 2016 Population Estimate from Michael Bauer Research, GmbH, derived from the Population Census from the Statistics Bureau, Ministry of Internal Affairs and Communications, Japan.

Modeling: Charlie Frye, Esri

MODELING GREEN INFRASTRUCTURE